Complex Networks and Dynamic Systems

Volume 4

Series Editor

Terry L. Friesz
Pennsylvania State University
University Park, PA, USA

More information about this series at http://www.springer.com/series/8854

Satish V. Ukkusuri • Chao Yang
Editors

Transportation Analytics in the Era of Big Data

 Springer

Editors
Satish V. Ukkusuri
School of Civil Engineering
Purdue University
West Lafayette, IN, USA

Chao Yang
Tongji University
Shanghai, China

ISSN 2195-724X   ISSN 2195-7258   (electronic)
Complex Networks and Dynamic Systems
ISBN 978-3-030-09344-0   ISBN 978-3-319-75862-6   (eBook)
https://doi.org/10.1007/978-3-319-75862-6

Printed on acid-free paper

This Springer imprint is published by the registered company Springer International Publishing AG part of Springer Nature.
The registered company address is: Gewerbestrasse 11, 6330 Cham, Switzerland

Preface

Historically transportation studies have primarily relied on survey-based approaches for understanding travel behavior and urban dynamics. The last few years has seen an explosion of "big data" techniques due to the rapid proliferation of various passive and mobile sensors in urban areas providing high resolution and large dimensional data in urban transportation systems. The book will focus on recent advances in the area of big data analytics and their applicability to solve various issues in transportation systems planning and operations. The core algorithmic approaches for big data techniques are based on machine learning (ML) methods. These have primarily been developed in computer science and engineering and have widely been in various engineering applications in the last few years [1–4]. These methods are relatively new in the transportation systems area and we estimate there are approximately around 55 scholars around the globe actively working in research, demonstrations, and applications pertaining to the various aspects of big data urban mobility modeling.

This book documents selected papers from the workshop on Big Data Analytics for Transportation Modeling, which was organized on July 16–17, 2016, at Tongji University by Professors Satish V. Ukkusuri and Chao Yang. This workshop brought together various international experts from academia, agencies, and industry to discuss emerging topics of interest in big data modeling. Various topics related to big data methods, data collection and curation, and applications for planning and operations in transportation systems were presented at the workshop. The organizers invited a selected set of participants to contribute chapters to this edited book. Finally, we selected about nine chapters for this book with authors from various universities in the world. Each book chapter was peer reviewed by at least two reviewers and was edited to fit with the main focus of the book. The book is divided into nine chapters.

Chapter 1 is based on utilizing tweet data for transportation planning applications. Particularly, the authors focused on estimating the localization of non-geotagged tweets using point of interest data and other spatial variables. This method provides rich information of both the geo-located tweets and estimates of aggregated location from non-geotagged tweets and demonstrated the usefulness for various transportation applications such as incident detection.

Chapter 2 explores the use of social media data from Twitter to complement other data sources to understand three transportation applications—traffic event detection, human mobility exploration, and trip purpose and demand forecasting. The chapter discusses how to leverage geolocation tweets to extract displacement of people and automatically extract topics of relevance for predicting event shifts.

Chapter 3 is written by some of the original collectors of taxi data in New York City from the Taxi and Limousine Commission (TLC). On the one hand, the chapter provides a historical perspective of the data collection initiative starting in 2004. On the other hand, transportation network providers have fought hard to prevent their data from being released to transportation regulators, frustrating their mission to implement policy, make and enforce regulations. The chapter highlights issues related to data accuracy, security, privacy, transparency, and compliance and the need for third-party independent institutions to audit and maintain this data.

Chapter 4 discusses a big data collection method using inertial measurement units (IMUs) to estimate vehicle path, detect traffic stops, and classify traffic-related events. The chapter discusses methods to estimate the vehicle trajectory from IMU and Bluetooth data and the mathematical problems that arise for an accurate estimation of such network-wide problems. The chapter discusses how to use this type of data to estimate traffic network states and at the same time maintaining the privacy of the data.

Chapter 5 discusses the data, methods, and applications of traffic source prediction, which may provide a new way to better understand the traffic congestion. The chapter discusses the "static driver source" that can be applied to urban planning and "dynamic driver source" that can support traffic management and control. The chapter also discusses "passenger source" in public transit system.

Chapter 6 discusses a sequential K-means clustering algorithm that utilizes smart card data to categorize Beijing subway stations, which are clustered into ten groups that are classified under three categories, i.e., employment-oriented, dual-peak, and residence-oriented stations. The chapter employs a geographically weighted regression model to determine the correlation effect between peak-hour passenger flow and land-use density. The findings of this chapter provide insightful information and theoretical foundations for rail transportation network design and operations management.

Chapter 7 discusses the rising concern for location privacy with the emergence of GPS capable mobile devices and the increasing demand of contextual services such as location-based services. The chapter discusses the method of geospatial analyses as an evaluation tool of the impact of noise-based algorithms on location data. The chapter identifies a threshold of noise settings so that privacy can be provided while geo-statistic inferences are not affected greatly.

Chapter 8 discusses PErsonal TRansport Advisor (PETRA) EU FP7 project, which is to develop an integrated platform to supply urban travelers with smart journey and activity advices, on a multi-modal network. The chapter discusses the architecture of PETRA platform and presents results obtained by applying PETRA to two different use cases, namely journey planning under uncertainty and smart tourism advisor with crowd balancing. The results of applying PETRA in two cities, Rome and Venice, are also presented.

Chapter 9 is on mobility pattern identification using mobile phone call record data (CRD) of 60 days obtained from Shenzhen, China. The chapter discusses representative features that were captured from each pattern in both weekday and weekend. The mobility pattern discussed in this chapter provides a new way to understand travel behavior, which plays a crucial role in urban planning and epidemic control.

In summary, the chapters in the book provide a rigorous understanding of big data methods, analytics, and applications to various transportation planning, operations, and control problems. Furthermore, some of the chapters discuss novel data collection methods and data privacy issues, which are important as we innovate in this area in the future.

Finally, we would like to acknowledge various people who have contributed to the completion of this book. We thank all authors who submitted their work for consideration. In addition, we thank the dozens of referees for their important work in reviewing the papers. We would also like to acknowledge the financial support provided for the workshop by Fundamental Research Funds of the Central Universities. We also thank Dr. Xiangdong Xu for support during the workshop preparation. Several students Xianyuan Zhan, Xinwu Qian, Fenfan Yan, and Yulaing Zhang also helped with the workshop organization.

West Lafayette, IN, USA   Satish V. Ukkusuri
Shanghai, China   Chao Yang
December 2017

References

1. P.S. Earle, D.C. Bowden, M. Guy, Twitter earthquake detection: Earthquake monitoring in a social world. Ann. Geophys. **54**, 708–715 (2011).
2. N. Naik, R. Raskar, C.A. Hidalgo, Cities are physical too: Using computer vision to measure the quality and impact of urban appearance. Am. Econ. Rev. **106**, 128–132 (2016). doi:10.1257/aer.p20161030
3. X. Zhan, X. Qian, S.V. Ukkusuri, A graph based approach to measuring the efficiency of urban taxi service system. IEEE Trans. Intell. Transp. Syst. **17**(9), 2479–2490 (2016).
4. X. Qian, S.V. Ukkusuri, Time of day pricing in taxi markets. IEEE Trans. Intell. Transp. Syst. **18**(6), 1610–1622 (2017).

Contents

9   Mobility Pattern Identification Based on Mobile Phone Data 217
Chao Yang, Yuliang Zhang, Satish V. Ukkusuri, and Rongrong Zhu

Chapter 1
Beyond Geotagged Tweets: Exploring the Geolocalisation of Tweets for Transportation Applications

Jorge David Gonzalez Paule, Yeran Sun, and Piyushimita (Vonu) Thakuriah

1.1 Introduction

There are currently several examples of the use of Twitter data in transportation analysis. These examples are in the areas of transportation operations [7, 9, 18, 28, 33, 44, 47] as well as planning [2, 20–22, 25, 52]. As knowledge of location is critical for many aspects of transportation research, virtually all such analyses utilise geotagged Twitter data, where precise location data is automatically generated once the user enables this feature. However, the number of tweets that are explicitly geotagged by users tends to be sparse (just 1–2% of the entire Twitter), so that the sample sizes are quite limited for real-time incident detection or event detection, particularly given the degree of noise and latency associated with tweets. Moreover, the spatial distribution of geotagged tweets is known to be non-representative of population patterns and Twitter users who publish geographical information have been found to be not representative of the wider Twitter population, causing problems with the use of geotagged Tweets for transportation planning and travel behaviour analysis.

This paper explores the extent to which the usefulness of Twitter data for transportation operations and planning can be improved by enhancing the geographic details of non-geotagged data. We call the process of estimating the locations of posts that are not associated with explicit coordinates "geolocalisation". Geolocalisation is an active research area in Twitter data analysis. We focus

J. D. G. Paule
School of Computing Science, University of Glasgow, Glasgow, UK
e-mail: j.gonzalez-paule.1@research.gla.ac.uk

Y. Sun · P. (Vonu) Thakuriah (✉)
Urban Big Data Centre, University of Glasgow, Glasgow, UK
e-mail: yeran.sun@glasgow.ac.uk; Piyushimita.Thakuriah@glasgow.ac.uk

© Springer International Publishing AG, part of Springer Nature 2019   1
S. V. Ukkusuri, C. Yang (eds.), *Transportation Analytics in the Era of Big Data*,
Complex Networks and Dynamic Systems 4,
https://doi.org/10.1007/978-3-319-75862-6_1

on one specific approach to enhancing the geographic details of non-geotagged tweets and consider two cases—first, automated incident detection and, second, the characteristics of location of people's activities—to empirically examine the extent to which geolocalisation improves transportation planning and operations, compared to what would be supported by geotagged data alone.

The paper is organised as follows: in Sect. 1.2, we discuss the main issues motivating our research approach, while in Sects. 1.3 and 1.4, we describe the data used and the geolocalisation methods adopted, respectively. Section 1.4.2 and its subsections present the approach undertaken for the comparative work on traffic incident detection using geotagged versus geolocalised Tweets whereas Sect. 1.4.3 looks at the spatial patterns of geotagged and geolocalised Tweets. Results are discussed in Sect. 1.5 and conclusions and recommendations for future research are drawn in Sect. 1.6.

1.2 Background

The range of data sources for transportation planning and operations in cities have grown rapidly, and the "*methods to manage and analyse such data have led to novel mobility services which may have the potential to lead to sustainable and socially interesting travel*" [48]. Particularly with the fast-growing utilisation of user-generated content (UGC) and crowd-sourced information for scientific research, several new approaches have emerged to study transportation patterns and travel behaviour. Popular social media such as Twitter, Instagram, Facebook and other sources can reveal not only historical travel patterns but also real-time traffic incidents and events. Yet the unstructured nature of the data and the level of noise involved in inferring knowledge can pose significant challenges to their routine use in transportation planning and operations.

One major area of interest in the transportation community is automated incident detection on roadways. This task depends on a wide variety of fixed sensors (inductive loop detection systems, CCTV) and moving-object sensors (probe vehicles, transit vehicles, cellphone users) and primarily covers the detection of events that disrupt efficient traffic operations. Typically in urban areas, roadways in higher functional classification tend to be instrumented by fixed sensors, while lower level arterial and side streets which are not as well equipped with monitoring with infrastructure-based sensors are monitored by moving objects and other ad-hoc sources. Separate (administrative data sources) provide information on crime and other incidents aboard public transportation vehicles and facilities. Safety researchers working with transportation data, on the other hand, work with data that are primarily collected by law enforcement officials at the site of transportation-related fatalities and serious injuries. Such sources of data together provide a comprehensive picture of the safe and efficient functioning of a city's transportation system; yet, these data sources are in disparate locations, available after different lengths of time and in heterogeneous formats, making it difficult to obtain a clear and complete perspective of the hazards and incidents in a city's transportation system in a timely manner.

Detecting small-scale road incidents using Twitter data has now been studied by many researchers but problems of detection rates, latency of detection, and other concerns are pertinent research issues. In the early days of using georeferenced tweets in detection of traffic events, only geotagged tweets are used due to high spatial granularity. Nevertheless, only about 1% of tweets are geotagged, and geotagged tweets are much more heterogeneously distributed than the overall population [17]. This means that an extremely limited number of georeferenced tweets are potentially useful in detection of traffic events with fine-grained occurrence locations.

1.2.1 Detection of Traffic Events

To detect traffic events by exploiting social media, some studies use both geotagged tweets and geolocalised tweets and find more tweets than using geotagged tweets alone [7, 9, 18, 28, 33, 44, 47]. Most of earlier studies on geolocalisation of tweets have limitations in either the precision of the spatial resolution recovered or the number of non-geotagged tweets for which location is estimated. Some studies geolocalise tweets at the nation or city level [13, 19, 27, 43] while others have retrieved a relatively small amount of fine-grained geolocalised tweets [14, 24, 31, 37, 38]. In this paper we try to retrieve a relatively large amount of street-level geolocalised tweets by using a new geolocalised approach. Then we use both these geolocalised tweets and the geotagged tweets to assess their comparative performance in the detection of traffic incidents in a metropolitan area.

1.2.2 Socio-Economic Characteristics of the Locations of GeoTagged Tweets

Regarding travel behaviour analysis for transportation planning studies, one important requirement is to understand the characteristics of the locations between which people travel (travel origins and destinations). The major sources of data to analyse these patterns are typically travel diaries from household travel surveys, and journey-to-work data that are collected by census organisations in many countries. Several studies have now reported the use of Twitter data to understand OD flow patterns. For example, Gao et al. [15], on the basis of 6.8 million geotagged tweets collected over a month from 110,868 users in the Greater Los Angeles area, evaluated the credibility of Twitter to estimate temporal mobility flows in comparison with the American Community Survey data. They concluded that their approach could be used to estimate aggregated mobility flows on weekdays. Kurcku et al. [29] found in the case of the New York metropolitan area that while flows between certain areas as detected from Twitter are a good match to ground truth data from the New York Metropolitan Transportation Council Regional Household Travel Survey, there were varying degrees of match to flows between other areas. Using Twitter to understand comprehensively understand OD flows is also challenging. Lee et al. [30] collected geotagged Twitter data over 2 days and identified 67,266 trips covering the entire Greater Los Angeles area; they commented on the large number of OD pairs with zero Twitter flows at the level of TAZs, requiring upward aggregation to the PUMA level to compare OD flows with the regional travel model.

1.2.3 Socio-Demographically Representativeness of GeoTagged Tweets

An assessment of the overall level of bias and variability in Twitter data is necessary to infer typical flow patterns in urban areas, as certain populations and their mobility patterns may be systematically under- or over-presented in Twitter data. These considerations have become tremendously important in the transportation planning literature and practice, in ascertaining, for example, access to jobs, health-care and other amenities for which there are social justice dimensions due to poverty and race, as well as age and gender. Concerns with travel quality of "Environmental Justice" (EJ) areas—areas with high low-income and minority populations—in particular have been important planning consideration. Geotagged tweets have been noted to be a non-representative sample of population patterns in general. Mislove et al. [36], for example, found that there is over-representation in geotagged tweets of populous counties and cities with large white populations in the USA, but underrepresentation of Hispanic populations in the Midwest and the southwest and of black populations in the south and the Midwest. Longley et al.[32] found in the UK (London) an over-representation of young males and white British users and an underrepresentation of middle-aged/older females, and of South Asian, West Indian and Chinese users. Hecht and Stephens [23] found, based on comparisons with census data, that urban areas have 2.7–3.5 times more geotagged tweet users than would be expected. Moreover, other authors have also compared geotagged Twitter activity against overall (including non-geotagged) Twitter activity, and have found (e.g., Sloan et al. [46]) that Twitter users who publish geographical information are socio-demographically not representative of the wider Twitter population (geotaggers being more likely to be male, slightly older, and to be small employers and in lower supervisory and technical occupations compared to non-geotaggers).

One approach to addressing the above problems is to increase the sample size of tweets with precisely known geographical location. Having a larger georeferenced sample would help in exploring issues of the overall representativeness, event coverage, latency and other concerns associated with geotagged data. Many approaches have been used to increase the range of tweets retrieved that enable the detection of incidents from tweets [7, 9, 18, 28]. A far from complete list of examples include querying through semantic enrichment of Twitter content [7, 28], increasing the sample of tweets through geolocalisation [18], and other approaches by means of which events and incidents can be detected. Regarding problems of inference, there is increased need (Thakuriah et al. [50]) as well as an interest in approaches to reducing sampling biases in online social networks and social media data [8, 10, 16, 40, 51]. While geolocalisation of non-geotagged tweets is an active area of research associated with the geospatial semantic Web and Geographic Information Retrieval, much of the work has focused on geolocalisation of users, or on geolocalisation of tweeting activity to fairly coarse geographical levels, whereas our work relates to street-level or even building-level geolocalisation. We will consider two different approaches to geolocalisation, as described in Sect. 1.4.

1.3   Data

In this work, we studied the Chicago metropolitan region. The area is defined by a bounding box with the following coordinates: $-86.8112, 42.4625, -88.4359, 41.2845$. To conduct our experiments, we collected Twitter data, traffic incident data and census data for a period of study of a month (July 2016).

1.3.1   Twitter Data

The Twitter dataset was collected from the Twitter Public Streaming API.[1] Spatial filtering can be applied in order to obtain georeferenced data. Geographical information is attached to tweets in two ways: (1) exact longitude and latitude if the GPS location reporting of the user device is activated (geotagged); and (2) as a suggested area (bounding box) from a list that can be extrapolated to a polygon, when sending a tweet (geobounded). In this work, we utilised geotagged and geobounded data for our experiments (see Table 1.1). The geobounded data provides a general location but not the spatial precision needed for the types of applications considered in this paper. Thus, we performed the geolocalisation on geobounded tweets. Since this work does not tackle geolocalisation per se but explores a practical way to go beyond geotagged data, we are using geobounded tweets for exemplification of the limitations of using geotagged tweets alone.

---

[1] https://dev.twitter.com/streaming/overview.

Table 1.1 Number of geotagged tweets and geobounded tweets collected (July 2016)

	Total tweets
Geotagged tweets	283,948
Geobounded tweets	2,357,360

1.3.2   Traffic Incidents

We gathered traffic incident data from the Bing Maps Traffic API.[2] The Bing Maps API provides real-time information about traffic disruptions happening in expressways and arterial roads. The data includes the start–end location as well as start–end time of the incident. In total, we collected 3182 traffic incidents in Chicago during the period of study (July 2016). The data also include several categories of incidents, however, in this study we focused on accidents. As a result, the final dataset is 1087 traffic accidents. Bing data incidents are primarily detected using inductive loops and CCTV, which are on the major roads and expressways. Other areas with minor roads are not reported, however Twitter data can provide coverage of such areas. The ultimate goal is to put together multiple data sources to obtain a more comprehensive picture.

1.3.3   Chicago Area SocioEconomic Data

Several data sources were used to understand the socioeconomic patterns in the Chicago metro area. One major source of data is the Chicago area Spatial Decision Support System constructed by the last author's research group [6, 49] that gives estimates of social, locational and behavioural factors at the level of census tract. Additionally, the Longitudinal Employer Household Dynamics (LEHD) synthetic data on Origin-Destination Employment Statistics (LODES) datasets[3] are used as well, which offer spatial distributions of jobs at the census block level, which we used aggregated to the census tract level.

1.4   Methodology

In this section, we first present the geolocalisation methods used in the paper. We then describe the process used to detect traffic incidents from both geotagged tweets and geolocalised tweets. This is followed by the approach used for an exploration of the spatial patterns of geotagged tweets and geolocalised tweets for the purposes of transportation planning applications by identifying spatial clusters of tweets and correlating tweets with population, job and other characteristics.

---

[2]https://msdn.microsoft.com/en-us/library/hh441725.aspx.
[3]https://lehd.ces.census.gov.

1.4.1   Geolocalisation of Tweets

Recently, geolocalisation of Twitter data has become an important yet challenging task. First efforts in the literature on geolocalising Twitter data addressed the localisation of Twitter users rather than individual tweets [3–5, 13, 19, 20, 26, 35]. However, despite these works can infer home and work locations of a Twitter user, they are unable to provide the current location of individual tweets.

Several works have proposed different approaches to address the problem of geolocalising individual tweets [11, 27, 37, 39, 43]. However, these studies achieved a coarse-level of granularity (i.e. country or city level) which remains insufficient for certain applications (e.g. traffic incident detection) that require more fine-grained geolocated data. Thus, geolocalising individual tweets at a fine-grained level has arisen as a new and challenging task that has been tackled recently [37, 38].

In this work we aim to obtain fine-grained geolocalised tweets in order to perform our analysis. We compared two different approaches to geolocalise tweets based on its content which utilise either external sources or twitter data in order to infer the locations of tweets based on their text. The first approach, called "POI", aims to match the presence of place names in the text with a Point Of Interest (POI) database. The second approach, called "Text-Based Similarity", retrieves the most likely location based on the similarity of the text of a non-geotagged tweet and the text of individual geotagged data. Both approaches predict the location of the tweet as a point with explicit longitude and latitude coordinates.

1.4.1.1   Geolocalisation Method 1: Point of Interest Matching

The first method exploits external sources to geolocalise tweets. Inspired by Schulz et al. [45], we implemented a method that matches the text of tweets with geospatial databases containing locally specific information. Some Twitter users explicitly mention the name of their location in their messages. Thus, by matching with a POI database, such names can be resolved to a point in space. POI databases consist of a list of place names, such as monuments, buildings or restaurants, that are associated with longitude and latitude coordinates.

We collected POI data from geospatial data sources that contain a higher number of place names that are unique for specific cities, and developed a method to perform a tweet-POI matching. The collected geospatial data sources are: Mapquest,[4] Bing[5] and Foursquare.[6] The POI matching extracts all possible n-grams from the text and match every n-gram with the POI database. An n-gram is a set of co-occurring words within a text in a given window (n). For example in the sentence *"I am in*

---

[4]https://developer.mapquest.com/.

[5]https://msdn.microsoft.com/en-us/library/hh478192.aspx.

[6]https://developer.foursquare.com/.

Chicago", the set of possible 3-grams is: {*"I am in"*, *"am in Chicago"*}. As a result, the method will output the POI that matches with the longest *n*-gram (up to 3-grams[7]) extracted from the text.

1.4.1.2   Geolocalisation Method 2: Text-Based Similarity Approach

Inspired by Gonzalez Paule et al. [38], we used information retrieval techniques to obtain the most similar geotagged tweet to a non-geotagged tweet based on the text. Information retrieval techniques in Twitter data have been exhaustively evaluated in the literature [42]. Given a set of documents, the classic information retrieval task aims to find the most relevant text documents to a given query based on the statistical similarity between the text of the documents and the text of the query. Therefore, given a set of geotagged tweets and a non-geotagged tweet for which we want to predict its location, we treat the text of the non-geotagged tweet as a query to search for the most similar documents within a collection of geotagged tweets. The intuition is, as users use Twitter to describe real-world events happening around them by means of their posts [1], then the similarity of the text also determines the proximity in geographical distance.

More specifically, the Vector Space Model (VSM) is applied using IDF (Inverse Document Frequency) statistic [34]. This model provides the best performance in tweet retrieval [42]. In VSM, each document and the query are represented as a single weighted vector in a multi-dimensional space. Each dimension in the space represents a word, and the weights are given by the IDF statistic. The IDF value for a word w in a collection with N documents is computed as follows:

$$idf_w = \log \frac{N}{df_w} \tag{1.1}$$

where df_w is the *document frequency* for word w, defined as the number of documents in the collection that contain the word. Therefore, for geolocalisation of tweets, we compute the IDF for each word in the collection of geotagged tweets. Next, each geotagged tweet and the non-geotagged tweet are represented as a vector containing the IDF values for each word in their text. Finally, the similarity between two documents is given by the distance between the vectors that represent them in the VSM. Thus, to retrieve the most similar geotagged tweet to the non-geotagged tweet, the cosine similarity between the vector of the geotagged tweet q and the vector of the non-geotagged tweet d is given by:

$$\text{cosine\_similarity}(q, d) = \frac{\mathbf{V}(q) \cdot \mathbf{V}(d)}{|\mathbf{V}(q)| \cdot |\mathbf{V}(d)|} \tag{1.2}$$

---

[7]3-grams is the best value to reduce matching ambiguity according to our experiments.

Table 1.2 Average error distance and recall for geolocation methods

	AVG error distance (km)	Recall (%)
POI	11.10	60.39
Text-based similarity	4.348	99.96

The IDF statistic ensures that rare words obtain high values whereas frequent words will obtain low values. This way, the most discriminative words within a spatial region are given a higher weight as they uniquely describe the area. The output of this process is a list of geotagged tweets ranked by the similarity score. Once the retrieval task is completed, the coordinates of the most similar geotagged tweet (top-1 in the list) are returned as the predicted location for the non-geotagged tweet.

1.4.1.3   Performance of Methods

Each of the two approaches described above has different levels of performance. We experimented over separate ground truth of 131,273 geotagged tweets from Chicago collected during March 2016 in order to evaluate the effectiveness of each approach. The geotagged tweets are utilised to perform predictions and then to compare the predicted location with the real location of the geotagged tweet. We utilised the first 3 weeks of our ground truth as a training set, and the last week as a testing set.

The following metrics are reported; **Average Error distance** (km), which is the distance on Earth (Haversine formula [41]) between the predicted location and the real coordinates of the tweet in our ground truth, and **Recall:** considered as the fraction of tweets in the test set we can geolocalise regardless of the distance error. Table 1.2 shows our experimental results. The "Text-Based Similarity" approach clearly outperforms the "POI" in error distance (11.10–4.348 km) and recall, finding a prediction for 99.96% of the evaluated tweets.

Additionally, we studied the distribution of the error produced by each of the geolocalisation methods. Figure 1.1 presents the Error Distance percentiles. As can be observed, the Text-Based Similarity approach ("Text-Based") performs better than the POI matching ("POI") for 75% of the data, improving by 7.306 km (11.378–4.072 km) the error with 50% of the geolocated tweets.

After evaluating both approaches in this way, we decided to utilise the Text-Based Similarity approach to perform the geolocalisation of non-geotagged tweets for the remainder of the analysis.

1.4.2   Application 1: Detection of Traffic Incidents

In this section we introduce our approach to identifying traffic-related tweets and link them with the incidents available from the Bing data in space and time. Firstly, we trained a text classifier using Naive Bayes algorithm to determine whether the content of the tweets is traffic related or not. Then we link the incidents with traffic-related tweets by its distance in space and time.

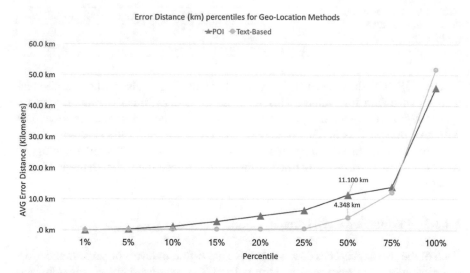

Fig. 1.1 Error Distance percentiles for geolocalisation methods. The *x*-axis shows the percentile. The *y*-axis shows the error distance

1.4.2.1 Finding Traffic-Related Tweets

As introduced before, the first task to tackle is to classify the tweets by its content, obtaining those that contain traffic-related information. To this end, we trained a Naive Bayes classifier, inspired by D'Andrea et al. [9], to determine whether the content of the tweets is traffic related or not, i.e., whether it contains traffic-specific words—such as car, crash, road, accident, etc. To this end, we collected a sample of 881 geotagged data from Chicago (March 2016) previously filtered by traffic-related keywords, and manually annotated them to train our classifier. The traffic-related keywords were selected manually following previous works and contain terms such as "accident", "car", "crash", etc. The annotation process resulted in 326 traffic-related tweets and 524 non-traffic-related tweets. The final evaluation showed that our model was able to correctly classify 81.45% of the tweets. The accuracy is similar with that in earlier studies [18]. For instance, [18] successfully identifies 82% of the tweets and [18] achieved 90.5% of accuracy.

1.4.2.2 Linking with Incidents

In the linkage phase, traffic-related tweets arc linked to traffic incidents that were detected in the same period of time. Our linkage strategy is based on a spatio-temporal matching criteria between tweets and incidents.

- Space Matching: We created a bounding box around the starting and ending locations of every incident. We considered 1–5 km range as our maximum threshold distance for every bounding box.
- Time Matching: Every traffic-related tweet posted between the starting time and the end time of the incident is considered.

1.4.3   Application 2: Exploring Spatial Patterns of Geotagged Tweets and Geolocalised Tweets for Transportation Planning Applications

We explore spatial patterns of geotagged tweets and geolocalised tweets, respectively, by (1) correlating tweets with population and other characteristics and (2) identifying spatial clusters of tweets. To properly associate tweets with population, employment and other socioeconomic factors, we first select tweets posted on workdays from both geotagged tweets and geolocalised tweets, and by removing tweets of users who made less than 30 tweets during the month to ensure the vast majority of tweets are likely to be made by local residents rather tourists. We conduct exploratory analysis of how tweets georeferenced by the two approaches differ according to locational and sociodemographic characteristics. By considering background population we also use the ratio of tweets to residents to identify clusters of high density tweets. Specifically, an improved AMOEBA (A Multidirectional Optimum Ecotope-Based) algorithm developed by Duque et al. [12] is used to identify clusters of high ratio of tweets to residents. Then we associate clusters with local built-up characteristics such as land use type and main non-residential buildings. As the residents and jobs data is available at the census tract level, we correlate tweets with population at the census tract level, and calculate ratio of tweets to residents at the census tract level.

1.4.3.1   Cluster Identification

As a spatially constrained clustering method, AMOEBA algorithm is applicable of classifying a large number of areas and identification of irregularly shaped clusters. Essentially, a region or ecotope is a spatially linked group of areas. A region can thus be defined as a spatially contiguous set of areas. The value of the G_i statistic is used to measure the level of clustering of an attribute x around an area. Suppose we run AMOEBA on a study region with N areas and an attribute x with elements x_i, indicating the value of x at area i. Let us denote this set of areas as M, and $\bar{x}$ and S as the mean and the standard deviation of the attribute x and let R be a subregion of M with n areas. Duque et al. [12] rewrite the formulation of G_R as follows:

$$G_R = \frac{\sum_{i \in R} x_i - n\overline{x}}{S\sqrt{\frac{Nn-n^2}{N-1}}} \qquad (1.3)$$

Accordingly, a positive (negative) and statistically significant value of G_i statistic indicates the presence of a cluster of high (low) values of attribute x around area i. Thus, AMOEBA identifies high-valued, or low-valued, ecotopes (regions) by looking for subsets of spatially connected areas with a high absolute value of the G_i statistic. There is only one parameter, i.e., the significance level threshold, that is required to run the AMOEBA algorithm. The significance level threshold was set to 0.01, meaning only clusters with a p-value less than 0.01 are considered. AMOEBA is implemented using the ClusterPy Python library of spatially constrained clustering algorithms.[8]

1.5   Results

In this paper, the study region is the Chicago Metropolitan region, including City of Chicago, with some areas in Illinois and a few others in Indiana. We first show the geolocalisation results (see Sect. 1.4.1). Then we present the experimental results of traffic incident detection by using geotagged tweets and geolocated tweets (Sect. 1.4.2). The results of the spatial patterns of geotagged tweets and geolocated tweets are then given in Sect. 1.4.3.

1.5.1   Geotagged Tweets and Geolocalised Tweets

We first collected 283,948 geotagged tweets within the study region for July 2016. By using the Text-Based Similarity geolocalised method aforementioned, we retrieved 1,747,938 geolocated tweets within the study region for July 2016. Furthermore, the Text Naive Bayes classifier method identified 2865 and 31,938 traffic-related tweets from the geotagged tweets and geolocated tweets, respectively. In total, there are 2,031,886 georeferenced tweets (geotagged + geolocalised). Of them 34,803 are traffic-related. Geolocalisation greatly increases the number of georeferenced tweets and also fine-grained traffic-related tweets (Table 1.3).

We also observed the spatial distribution of geotagged and geolocated tweets. As can be seen in Fig. 1.2, geolocated tweets are more heterogeneously distributed over the region than the geotagged tweets, which are more concentrated in CBD area of the City of Chicago and to the Near North part of the city. There are some other areas such as in the northern suburbs where also we see groupings of geotagged

---

[8]www.rise-group.org/risem/clusterpy.

Table 1.3 Statistics on georeferenced tweets and traffic-related tweets

	Total tweets	Traffic related
Geotagged tweets	283,948	2865
Geolocated tweets	1,747,938	31,938
Total georeferenced tweets	2,031,886	34,803

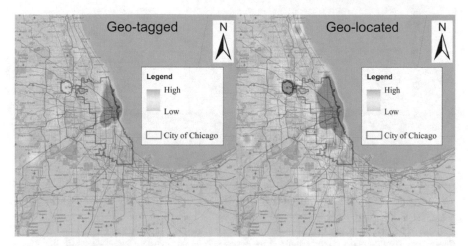

Fig. 1.2 Heatmaps of all geotagged tweets and all geolocalised tweets (Heatmaps are created by using Kernel Density Function)

tweets. The geolocalised tweets, on the other hand, can also be prominently seen in many areas in the western and southern parts of the metropolitan area where there is little geotagged activity, giving strong evidence that twitter activity in these areas is not non-existent—they may simply not be geotagged.

1.5.2   Application 1: Detection of Traffic Incidents

Figure 1.3 shows the incident detection rate for geotagged, geolocated and the combination of geotagged and geolocated Twitter data. We compared the percentage of incidents covered by when utilising different spatial ranges. As it can be seen, the number of detected incidents is significantly increased when utilising geolocalising methods. The percentage of incidents detected when adding extra geolocated tweets to the geotagged data also improves: increase of 7.72% within 1 km distance (6.53–14.25%) to 33.85% of increase within 5 km of distance (17.93–51.79%). As expected, the detection rate increases rapidly along with the spatial radius. This is due not only to the larger area covered but also by the number of tweets geolocated at a fine-grained level and nearby to its real location. The geolocation methods evaluation showed that the mean distance error for the Text-Based Similarity approach is 4.346 km.

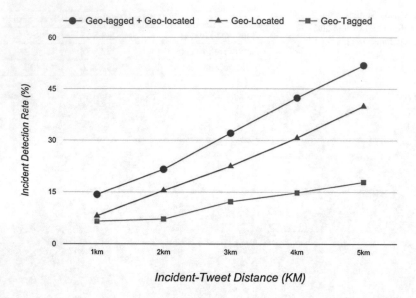

Fig. 1.3 Incident detection rate at different distances for Geotagged data, Geolocated data and aggregated (Geotagged + Geolocated). The y-axis shows the incident rate (%) while the x-axis shows the distance threshold in kilometres between the tweet and the incident

Additionally, we also discovered tweets that are potentially reporting traffic incidents in areas that are not reported by other data sources such as Bing. Heatmaps of geotagged traffic-related tweets and geolocalised traffic-related tweets that could not be linked to Bing traffic incidents are presented in Fig. 1.4, from where the location of these unlinked incidents not covered by our incident dataset can be seen. One simple explanation is that these are locations of where people tweeted about the incident from, not the actual location of the incident—and hence further investigation is merited—although the level of clustering of these unlinked incidents points to this spatio-temporal asymmetry being only part of the explanation. A second possible explanation is that the Bing traffic incident dataset only contains motorised traffic incidents that occur in expressways and the major arterial roads, whereas Twitter conversation in the unlinked tweets is about more minor incidents in more local roads. It is also well known the incidents and crashes in many deprived Environmental Justice areas in particular are significantly underreported [6], and the heatmaps of geolocalised unlinked tweets in Fig. 1.4 certainly indicate activity in such EJ areas in the City of Chicago. While further analysis is merited, it is possible to hypothesise at this stage that traffic-related content in Twitter data covers a larger area and, therefore, incidents that are not covered by traditional detection infrastructure.

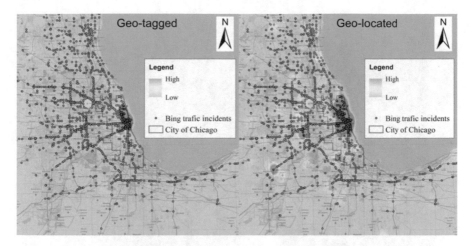

Fig. 1.4 Heatmaps of geotagged traffic-related tweets and geolocated traffic-related tweets not linked to Bing traffic incidents (Heatmaps are created by using Kernel Density Function)

Table 1.4 Correlations of tweets and residents or jobs

	Correlations	
Variable	#Geotagged tweets	#Geolocated tweets
Number of residents	0.11	0.09
Number of jobs	0.91	0.35

1.5.3   Application 2: Spatial Patterns of Geotagged Tweets and Geolocalised Tweets

We use Pearson's correlation coefficient to measure the correlations of numbers of residents and jobs with numbers of geotagged and geolocalised tweets, shown in Table 1.4. The number of geotagged tweets is more strongly correlated to the number of jobs than the number of geolocalised tweets. This indicates that geotagged tweets are spatially associated with workplaces where there may be better access to the Internet and where working populations spend most of the time during the day, and are more easily to post tweets.

t-Tests given in Table 1.5 further confirm that census tracts with higher numbers of jobs have a statistically higher number of geotagged tweets compared to tracts with lower number of jobs. The same trend can generally be seen for the geolocalised tweets. In contrast, whereas tracts with lower levels of residential population density have a statistically higher average number of geotagged tweets, compared to areas with higher levels of population density, this is not the case for geolocalised tweets. This implies that geolocalised tweets are more likely to be evenly spread across high and low levels of population density levels giving a more even representation of underlying activity patterns.

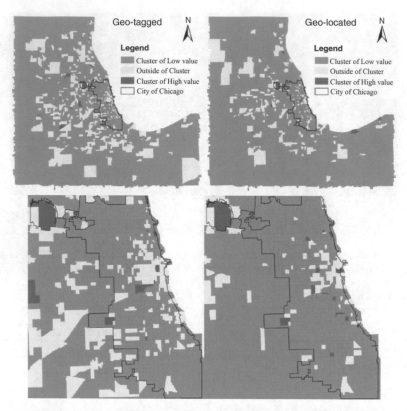

Fig. 1.5 Clusters of high ratio of tweets to residents for geotagged data and geolocated data respectively

The smoothing of the density of tweets by geolocalisation across the metropolitan area can be visualised by considering clustering of geotagged versus geolocalised tweets. The AMOEBA algorithm identifies statistically significant clusters of high value and clusters of low value. Figure 1.5 shows the maps of the cluster of high ratio of tweets to residents (RTR). In these maps, clusters of high value and low value represent cluster of high RTR and low RTR, respectively. We further associate clusters of high RTR with local built-up characteristics such as main non-residential buildings by overlapping the clusters and basemap such as GoogleMap and OpenStreetMap. Accordingly, clusters of high RTR for geotagged tweets mainly surround airports, shopping malls and institutions including universities, colleges and schools; whilst clusters of high RTR for geolocated tweets mainly surround not only major transportation hubs and institutions but also stadiums and sporting venues, as well as purely residential areas scattered throughout the metro area. This implies that compared to geotagged tweets, geolocated tweets have a larger portion that is related to all types of activities, not only to social and work activity locations as in the case of geotagged tweets.

Table 1.5 t-Statistics on geotagged and geolocalised tweets by socioeconomic factors

Factor	Means			
	Geotagged	t	Geolocalised	t
Jobs = Low	12.86	4.78*	181.1	4.70*
Jobs = High	69.92		659.3	
Pop. Sq. Mile = Low	48.23	−1.98**	416	0.13
Pop. Sq. Mile = High	27.58		429.4	
EJ = No	56.86	5.20*	481.1	0.82
EJ = Yes	16.09		376.6	
Young = Low	43.61	−0.75	428.4	−0.24
Young = High	36.67		403	

*Significant at 0.01
**Significant at 0.05

The lowered levels of representativeness of the geotagged tweets compared to the geolocalised tweets can be seen further when we consider EJ areas, which as discussed previously represent areas with high presence of low-income and minority populations. We followed the definition of EJ areas given in Cottrill and Thakuriah [6]. Due to issues of digital divide, it may be expected that the level of social media activity is potentially lower in EJ areas. Table 1.5 also shows that the number of both geotagged and geolocalised tweets is not significantly different in areas with a high levels of young people (less than 30 years of age). However, these differences are reduced with the geolocalisation, as the numbers of tweets geolocated in EJ areas increase in relation to the numbers for non-EJ areas. This is a useful development towards using Twitter data in transportation planning because geolocalisation enhances access to social media activity in these deprived areas, which would otherwise be quite masked if using geotagged data only.

Finally, Table 1.5 also shows that the number of both geotagged and geolocalised tweets is not significantly different in areas with a high levels of young people (less than 30 years of age). In the case of both types of Twitter localisation, areas with lower numbers of young people as resident populations have higher levels of Twitter activity. This is in the direction of the findings by Sloan et al. [46]) mentioned in Sect. 1.2.

1.6   Conclusions

The major motivation of this paper was to evaluate benefits of enhanced geolocalisation of Twitter data. We demonstrated that combining geolocalised tweets and geotagged tweets enables a higher degree of performance in detection of traffic incidents than using geotagged tweets alone. Geolocalisation of tweets enables detection of more traffic incidents although the occurrence locations might be not as accurate as using geotagged tweets alone. Although further analysis is warranted, we found geolocalised incidents not available in the incident records, particularly in areas where there has been historically underreporting of road crashes and incidents. We also found that using geolocalised tweets allows discovery of social media activity throughout the metropolitan area, including deprived EJ areas. Enhanced geolocalisation therefore makes Twitter data better suited to transportation planning applications due to a higher degree of representativeness of a larger set of activity and area characteristics, in contrast to what is available from the publicly available geotagged data.

The approach used in the paper has several limitations that merit future consideration. First, we only use tweets for 1 month as experimental data. There are potential biases and issues of seasonality and sparsity. Second, we take account of only one traffic event: traffic incidents. Other types of traffic events including road congestion and transit delay are also of concern in transportation planning and operations. Third, the approach used for the detection of traffic incidents in this study is simple and subject to historical ground-truth data. In the detection method, spatio-temporal ranges used to detect traffic incidents centre known traffic incidents, although as shown, georeferenced tweets might indicate some traffic events that not covered by the ground-truth data used. Fourth, the mean distance error for the geolocalisation method is 4–5 km, street-level usage of geolocated tweets is at high risk of placement error. More caution is required when using geolocated tweets in large-scale applications. Aggregate geolocated tweets into geographic areas such as census tracts would increase influence of tweets' locational errors on transportation planning applications. Finally, the comparison with socioeconomic patterns utilises spatially aggregated census and other data at the level census tracts, due to which a true comparison of the validity of the precise geolocalisation was not possible.

Future research will involve collecting a larger set of tweets for a longer time periods such as a few months to repeat the experiments in this study to evaluate sensitivity of the approach. Second, this study is an exploratory analysis dedicated to evaluating the benefit of enhanced geolocalisation in the detection of traffic events and socioeconomic patterns. Future work will also involve comparing detection of different types of traffic events to understand spatial, temporal and content coverage of geolocated tweets for different types of traffic-related information. For example, there is a need for modelling approaches to detect events independent of spatio-temporal ranges of known traffic incidents. This might be useful in detection of very small-scale historical or real-time traffic incidents from tweets. Finally, much analysis is needed to evaluate the benefit of using different types of geolocalised data in inferring changing socioeconomic patterns in various parts of the city with difference characteristics.

Acknowledgements The research was supported by European Commission FP7 Grant No 632075 and the Research Council of UK's Economic and Social Research Council Grant No ES/L011921/1. The authors would also like to express gratitude to Dr. Yashar Moshfeghi and Professor Joemon M. Jose for their assistance in providing the methods and data utilised in this work.

References

1. M.A. Abbasi, S.K. Chai, H. Liu, K. Sagoo, Real-world behavior analysis through a social media lens, in *Proceedings of the 5th International Conference on Social Computing, Behavioral-Cultural Modeling and Prediction, SBP'12* (Springer, Berlin, 2012), pp. 18–26
2. F. Alesiani, K. Gkiotsalitis, R. Baldessari, A probabilistic activity model for predicting the mobility patterns of homogeneous social groups based on social network data, in *Transportation Research Board 93rd Annual Meeting, 14-1033* (2014)
3. H.w. Chang, D. Lee, M. Eltaher, J. Lee, @phillies tweeting from philly? Predicting twitter user locations with spatial word usage, in *Proceedings of the 2012 International Conference on Advances in Social Networks Analysis and Mining, ASONAM 2012* (IEEE Computer Society, Washington, 2012), pp. 111–118. https://doi.org/10.1109/ASONAM.2012.29
4. Z. Cheng, J. Caverlee, K. Lee, You are where you tweet: a content-based approach to geo-locating twitter users, in *Proceedings of the 19th ACM International Conference on Information and Knowledge Management* (ACM, New York, 2010), pp. 759–768
5. R. Compton, D. Jurgens, D. Allen, Geotagging one hundred million twitter accounts with total variation minimization, in *2014 IEEE International Conference on Big Data (Big Data)* (IEEE, Piscataway, 2014), pp. 393–401
6. C.D. Cottrill, P.V. Thakuriah, Evaluating pedestrian crashes in areas with high low-income or minority populations. Accid. Anal. Prev. **42**(6), 1718–1728 (2010)
7. J. Cui, R. Fu, C. Dong, Z. Zhang, Extraction of traffic information from social media interactions: methods and experiments, in *2014 IEEE 17th International Conference on Intelligent Transportation Systems (ITSC)* (IEEE, Piscataway, 2014), pp. 1549–1554
8. A. Culotta, Reducing sampling bias in social media data for county health inference, in *Joint Statistical Meetings Proceedings* (2014), pp. 1–12
9. E. D'Andrea, P. Ducange, B. Lazzerini, F. Marcelloni, Real-time detection of traffic from twitter stream analysis. IEEE Trans. Intell. Transp. Syst. **16**(4), 2269–2283 (2015)
10. O. Dekel, O. Shamir, Vox populi: collecting high-quality labels from a crowd, in *COLT* (2009)
11. M. Dredze, M.J. Paul, S. Bergsma, H. Tran, Carmen: a twitter geolocation system with applications to public health, in *Proceedings of the AAAI Workshop on Expanding the Boundaries of Health Informatics Using Artificial Intelligence*, Palo Alto, California (2013)
12. J.C. Duque, J. Aldstadt, E. Velasquez, J.L. Franco, A. Betancourt, A computationally efficient method for delineating irregularly shaped spatial clusters. J. Geogr. Syst. **13**(4), 355–372 (2011)
13. J. Eisenstein, B. O'Connor, N.A. Smith, E.P. Xing, A latent variable model for geographic lexical variation, in *Proceedings of the 2010 Conference on Empirical Methods in Natural Language Processing, EMNLP '10* (Association for Computational Linguistics, Stroudsburg, 2010), pp. 1277–1287. http://dl.acm.org/citation.cfm?id=1870658.1870782
14. D. Flatow, M. Naaman, K.E. Xie, Y. Volkovich, Y. Kanza, On the accuracy of hyper-local geotagging of social media content, in *Proceedings of the Eighth ACM International Conference on Web Search and Data Mining* (ACM, New York, 2015), pp. 127–136
15. S. Gao, J.A. Yang, B. Yan, Y. Hu, K. Janowicz, G. McKenzie, Detecting origin-destination mobility flows from geotagged tweets in greater Los Angeles area, in *Eighth International Conference on Geographic Information Science, GIScience'14* (2014)
16. M. Gjoka, M. Kurant, C.T. Butts, A. Markopoulou, Walking in facebook: a case study of unbiased sampling of osns, in *2010 Proceedings IEEE Infocom* (IEEE, New York, 2010), pp. 1–9
17. M. Graham, S.A. Hale, D. Gaffney, Where in the world are you? Geolocation and language identification in twitter. Prof. Geogr. **66**(4), 568–578 (2014)
18. Y. Gu, Z.S. Qian, F. Chen, From twitter to detector: real-time traffic incident detection using social media data. Transp. Res. Part C Emerg. Technol. **67**, 321–342 (2016)

19. B. Han, P. Cook, A stacking-based approach to twitter user geolocation prediction, in *Proceedings of the 51st Annual Meeting of the Association for Computational Linguistics (ACL 2013): System Demonstrations* (2013), pp. 7–12

20. B. Han, P. Cook, T. Baldwin, Text-based twitter user geolocation prediction. J. Artif. Intell. Res. **49**, 451–500 (2014)

21. S. Hasan, S.V. Ukkusuri, Urban activity pattern classification using topic models from online geo-location data. Transp. Res. Part C Emerg. Technol. **44**, 363–381 (2014)

22. S. Hasan, S.V. Ukkusuri, Location contexts of user check-ins to model urban geo life-style patterns. PLoS One **10**(5), e0124819 (2015)

23. B. Hecht, M. Stephens, A tale of cities: urban biases in volunteered geographic information, in *International Conference on Weblogs and Social Media*, vol. 14 (2014), pp. 197–205

24. Z. Ji, A. Sun, G. Cong, J. Han, Joint recognition and linking of fine-grained locations from tweets, in *Proceedings of the 25th International Conference on World Wide Web* (International World Wide Web Conferences Steering Committee, Republic and Canton of Geneva, 2016), pp. 1271–1281

25. P. Jin, M. Cebelak, F. Yang, J. Zhang, C. Walton, B. Ran, Location-based social networking data: exploration into use of doubly constrained gravity model for origin-destination estimation. Transp. Res. Rec. J. Transp. Res. Board **2430**, 72–82 (2014)

26. D. Jurgens, That's what friends are for: inferring location in online social media platforms based on social relationships, in *International Conference on Weblogs and Social Media*, vol. 13 (2013), pp. 273–282

27. S. Kinsella, V. Murdock, N. O'Hare, I'm eating a sandwich in glasgow: modeling locations with tweets, in *Proceedings of the 3rd International Workshop on Search and Mining User-Generated Contents* (ACM, New York, 2011), pp. 61–68

28. R. Kosala, E. Adi, et al., Harvesting real time traffic information from twitter. Procedia Eng. **50**, 1–11 (2012)

29. A. Kurkcu, K. Ozbay, E.F. Morgul, Evaluating the usability of geo-located twitter as a tool for human activity and mobility patterns: a case study for New York city, in *Transportation Research Board 95th Annual Meeting*, 16-3901 (2016)

30. J.H. Lee, S. Gao, K. Janowicz, K.G. Goulias, Can twitter data be used to validate travel demand models? in *IATBR 2015-WINDSOR* (2015)

31. C. Li, A. Sun, Fine-grained location extraction from tweets with temporal awareness, in *Proceedings of the 37th International ACM SIGIR Conference on Research and Development in Information Retrieval* (ACM, New York, 2014), pp. 43–52

32. P.A. Longley, M. Adnan, G. Lansley, The geotemporal demographics of twitter usage. Environ. Plan. A **47**(2), 465–484 (2015)

33. E. Mai, R. Hranac, Twitter interactions as a data source for transportation incidents, in *Proceedings of the Transportation Research Board 92nd Annual Meeting*, 13-1636 (2013)

34. C.D. Manning, P. Raghavan, H. Schütze, et al., *Introduction to Information Retrieval*, vol. 1 (Cambridge University Press, Cambridge, 2008)

35. J. McGee, J. Caverlee, Z. Cheng, Location prediction in social media based on tie strength, in *Proceedings of the 22nd ACM International Conference on Information and Knowledge Management* (ACM, New York, 2013), pp. 459–468

36. A. Mislove, S. Lehmann, Y.Y. Ahn, J.P. Onnela, J.N. Rosenquist, Understanding the demographics of twitter users, in *5th International Conference on Weblogs and Social Media*, vol. 11 (2011)

37. P. Paraskevopoulos, T. Palpanas, Where has this tweet come from? Fast and fine-grained geolocalization of non-geotagged tweets. Soc. Netw. Anal. Min. **6**(1), 89 (2016)

38. J.D.G. Paule, Y. Moshfeghi, J.M. Jose, P. Thakuriah, On fine-grained geo-localisation of tweets, in *Proceedings of the 2017 ACM International Conference on the Theory of Information Retrieval, ICTIR '17* (ACM, New York, 2017). https://doi.org/10.1145/3121050.3121104

39. R. Priedhorsky, A. Culotta, S.Y. Del Valle, Inferring the origin locations of tweets with quantitative confidence, in *Proceedings of the 17th ACM Conference on Computer Supported Cooperative Work and Social Computing* (ACM, New York, 2014), pp. 1523–1536

40. V.C. Raykar, S. Yu, L.H. Zhao, G.H. Valadez, C. Florin, L. Bogoni, L. Moy, Learning from crowds. J. Mach. Learn. Res. **11**(4), 1297–1322 (2010)
41. C.C. Robusto, The cosine-haversine formula. Am. Math. Mon. **64**(1), 38–40 (1957)
42. J.A. Rodriguez Perez, J.M. Jose, On microblog dimensionality and informativeness: exploiting microblogs' structure and dimensions for ad-hoc retrieval, in *Proceedings of the 2015 International Conference on The Theory of Information Retrieval, ICTIR '15* (ACM, New York, 2015), pp. 211–220. https://doi.org/10.1145/2808194.2809466
43. A. Schulz, A. Hadjakos, H. Paulheim, J. Nachtwey, M. Mühlhäuser, A multi-indicator approach for geolocalization of tweets, in *International Conference on Weblogs and Social Media* (2013)
44. A. Schulz, P. Ristoski, H. Paulheim, I see a car crash: real-time detection of small scale incidents in microblogs, in *The Semantic Web: ESWC 2013 Satellite Events* (Springer, Berlin, 2013), pp. 22–33
45. A. Schulz, B. Schmidt, T. Strufe, Small-scale incident detection based on microposts, in *Proceedings of the 26th ACM Conference on Hypertext and Social Media* (ACM, New York, 2015), pp. 3–12
46. L. Sloan, J. Morgan, Who tweets with their location? Understanding the relationship between demographic characteristics and the use of geoservices and geotagging on twitter. PLoS One **10**(11), e0142209 (2015)
47. E. Steiger, T. Ellersiek, A. Zipf, Explorative public transport flow analysis from uncertain social media data, in *Proceedings of the 3rd ACM SIGSPATIAL International Workshop on Crowdsourced and Volunteered Geographic Information, GeoCrowd '14* (ACM, New York, 2014), pp. 1–7. https://doi.org/10.1145/2676440.2676444
48. P. Thakuriah, D.G. Geers, *Transportation and Information: Trends in Technology and Policy* (Springer, Berlin, 2013)
49. P. Thakuriah, P. Metaxatos, J. Lin, E. Jensen, An examination of factors affecting propensities to use bicycle and pedestrian facilities in suburban locations. Transp. Res. Part D Transp. Environ **17**(4), 341–348 (2012)
50. P. Thakuriah, N. Tilahun, M. Zellner, Big data and urban informatics: innovations and challenges to urban planning and knowledge discovery, in *Seeing Cities Through Big Data: Research Methods and Applications in Urban Informatics*, chap. 10, ed. by P. Thakuriah, N. Tilahun, M. Zellner (Springer, New York, 2016), pp. 11–45
51. F.L. Wauthier, M.I. Jordan, Bayesian bias mitigation for crowdsourcing, in *Advances in Neural Information Processing Systems* (2011), pp. 1800–1808
52. F. Yang, P.J. Jin, X. Wan, R. Li, B. Ran, Dynamic origin-destination travel demand estimation using location based social networking data, in *Transportation Research Board 93rd Annual Meeting*, 14-5509 (2014)

Chapter 2
Social Media in Transportation Research and Promising Applications

Zhenhua Zhang and Qing He

2.1 Social Media Explosion

Recent years have witnessed the newly emerging techs of social media in shaping our lives. The evolution of Information and Communication Technology (ICT) comes with the increasing coverage of smartphone and other mobile devices. Such ICT-based technologies allow individuals, companies, NGOs, governments, and other organizations to view, create, and share information, ideas, and career interests [1]. Social media exists in different forms including the traditional news, media, or websites and has evolved with the transformation of communication technologies. Social media, a kind of "We Media," converts the traditional one-directional news feed from large media organizations to the individuals, into bidirectional information sharing where everyone becomes a news center. Researchers can thus acquire the wide-range information from the massive majority of people. The well-known social media websites include Facebook, WeChat, Tumblr, Instagram, Twitter, Baidu Tieba, Pinterest, LinkedIn, Google+, YouTube, Viber, and Snapchat. These social media tools and websites bring promising applications using social media to solve transportation problems.

The explosion and popularity of social media can be manifested by the monthly active users acquired from DMR websites [2] and Statista.com (Table 2.1).

Social media technologies take many different forms including blogs, business networks, enterprise social networks, forums, microblogs, photo sharing, prod-

Z. Zhang
Iowa State University, Ames, IA, USA
e-mail: zhenhuaz@iastate.edu

Q. He (✉)
The State University of New York, University at Buffalo, Buffalo, NY, USA
e-mail: qinghe@buffalo.edu

© Springer International Publishing AG, part of Springer Nature 2019
S. V. Ukkusuri, C. Yang (eds.), *Transportation Analytics in the Era of Big Data*,
Complex Networks and Dynamic Systems 4,
https://doi.org/10.1007/978-3-319-75862-6_2

Table 2.1 Basic statistics of some well-known social media

Social media name	Monthly active users (M)	Release date
Facebook	1650 (3/31/2016)	2004
WeChat	700 (4/31/2016)	2011
Tumblr	555 (9/1/2016)	2007
Instagram	500 (6/21/2016)	2010
Twitter	310 (5/1/2016)	2006
Baidu Tieba	657 (2/5/2016)	2003
Pinterest	100 (9/18/2015)[a]	2010
LinkedIn	106 (4/28/2016)	2002
Google+	540 (2/1/2016)	2011
YouTube	1000 (4/21/2014)	2005

[a]This is the total number of Pinterest users

ucts/services review, social bookmarking, social gaming, social networks, video sharing, and virtual worlds [3]. Each has its own specialized focus. For instance, Twitter and Sina Weibo focus on sharing short message streams; LinkedIn is mainly for business networks and career purposes; Square shares the dining and recreational places; Baidu Tieba is built on the forum website and focuses on various social topics; Wikipedia is a knowledge-sharing website; Facebook builds a general platform for social networks and a wide range of online applications. Table 2.2 summarizes the focuses of social media and the representative social media websites and tools in North America. The focuses are subject to change and involve with times passing by, and new forms of social media are coming forth continuously [4].

Despite the different types of social media, social media has common features which make them useful for transportation applications. The first one is a social reflection in which social media reflects the social events such as traffic accident studies, traffic jam, etc. The second feature is that data collection from social media is usually cheaper than traditional data acquisition, and some of the companies provide open API for the purpose. The third feature is that social media usually have multi-topics and this will potentially broaden the horizon of the transportation-related studies. Table 2.3 listed parts of the applications on social media, and more related studies are still undergoing.

Information from Twitter may be biased, and the representativeness of the Twitter users can reveal some important details of certain groups of people. Although we cannot directly interview the Twitter users in our collected datasets, there is still some open survey or investigation estimating their demographics. Table 2.4 lists education, age, gender, and income distributions of Twitter users from comScore [36], Pew Research Center [37], and statista.com [38, 39].

This chapter will exemplify the promising transportation-related applications of social media in two categories. The first category of studies is built on the location and time data from social media and explores features of human mobility, travel behavior, etc., under certain circumstances; the second category mainly focuses on the text interpretation and employs the state-of-the-art language modeling techniques to extract useful transportation-related information.

Table 2.2 Focuses of social media and their representatives

Focus	Social media website and tool
Live casting	Actors Access, Backstage, Actors Equity Casting Call, Playbill, SAG Indie, Now Casting, Casting Networks, NYCastings.com, Mandy.com, Craigslist
Virtual world	ourWorld, Wizard 101, Woozworld, Virtual Families, Second Life, IMVU, Habbo Hotel, Smeet, Meez, SmallWorlds
Wiki	Wikipedia, Wikitravel, wikiHow, Wikibooks, CookBookWik, WikiSummaries, wikimapia, Wiktionary, Uncyclopedia, ProductWiki, LyricWiki, Wikicars
Music	Pandora, Yahoo! Music, Google Play, SoundCloud, Spotify, MySpace, TuneIn, Last.fm, iHeart, AllMusic, Jango, Radio, Songza, Live365, Slacker
Event sharing	Beadwork, Folio, Forbes, HOW, Dwell, Farm Progress, WineMaker, Entrepreneur, Wired
Document sharing	ISSUE, SlideShare, Scribd, Box.net, DocStock, Calameo, Zoho, Keep and Share, Free eBooks, 4shared, Author Stream, Wattpad, YUDU, Wuala, Div Share, ADrive
CRM (customer relationship management)	Salesforce, Microsoft Dynamics CRM, Oracle Sales Cloud, SugarCRM, Workbooks CRM, Insightly, Nimble, Zoho CRM, NetSuite CRM, Veeva CRM
Video	YouTube, Vimeo, Yahoo! Screen, Dailymotion, Hulu, Vube, Twitch, LiveLeak, Vine, Ustream, Break, TV.com, Metacafe
Reviews and ratings	Amazon Customer Reviews, Angie's List, Choice, Trustpilot, TestFreaks, Which?, ConsumerReports, TripAdvisor, Yelp, Google My Business, Yahoo! Local Listings, G2 Crowd, TrustRadius, Salesforce AppExchange, Better Business Bureau, Glassdoor
Business relationships	AngelList, Beyond, Black Business Women Online, Data.com Connect, E.Factor, Gadball, Gust, LinkedIn, Meetup, Networking for Professionals, Opportunity, PartnerUp, PerfectBusiness, Plaxo, Quibb, Ryze, StartupNation, Upspring, Viadeo
Dashboards	toutapp, Lancaster Bingo Company, A flat design dashboard, Yet more flat design, fitbit, Patient records, Sprout Social, Nektar Dashboard, Wufoo, Dashboard Analytics, Cranium Dashboard, Start Admin, Fox Metrics

(continued)

Table 2.2 (continued)

Focus	Social media website and tool
General networking	Facebook, Hi5, myspace.com, renren, Bebo, PerfSpot
Discussion boards and forums	phpBB, Simple Machines Forum, ZetaBoards, bbPress, Vanilla Forums, PunBB, fluxBB, PlushForums, Phorum, MyBB, miniBB
DIY and custom	Ana-White.com, Shanty-2-Chic, Jay's Custom Creations, ArtofManliness.com, Instructables
Blogs	Blog.com, Blogger.com, Medium.com, Penzu.com, Squarespace.com, Svbtle.com, Tumblr.com, Webs.com, Weebly.com, Wix.com, WordPress.com
Microblog	Twitter, Friend Feed, Tumblr, Plurk, Pinterest, Flattr, Dipity, Yammer, MeetMe, Plerb
(Q&A) Questions and answers	Quora, Mind the Book, Amazon's askville, Yahoo! Answers, Stack Overflow, Super User, LinkedIn Answers, Answers.com, Hacker News' Ask Section, LawPivot
Social commerce	Pinterest, Shopee, Lyst, Soldsie, Kickstarter
Pictures	Instagram, Imgur, Flickr, Photobucket, DeviantArt, Shutterfly, TinyPic, WeHeartIt, ImageShack, ImageVenue, SmugMug

2.2 Applications Based on Social Media

2.2.1 Traffic Event Detection

As tweets are able to describe what is happening on the scene site and the tweeting locations may be quite near the scene site, the tweet content analysis is usually the priority in most of the studies. Over the past decades, the online texts posted by social media have been validated useful to broadcast major events such as natural disasters [14, 40], bird flu [41], politic events [42], etc. The traffic event detection also arouses increasing attentions: Mai et al. [43] compared incident records with Twitter messages and proved the potentials of Twitter as a supplemental traffic measurement. Schulz et al. [16] used microblogs to detect the small-scale incidents. Gal-Tzur et al. [44] conducted a corridor study on the correlation between a tweet and traffic jam. Gu et al. [45] combined the data sources from Twitter, incident records, HERE, etc., and employed the Naïve-Bayes classification to detect five major incident types. D'Andrea et al. [46] compared accuracies and precisions of different regression models including Naïve-Bayes, Support Vector Machine, Artificial Neural Network, Decision Tree in detecting traffic incidents from Twitter stream. Our preliminary examinations also demonstrate the potential of Twitter in delivering the accident-related information (Table 2.5).

To automatically detect events from social media, there are several challenges to be addressed. Taking traffic accident detection as an example: First, as compared to events that arouse enormous public attentions such as sporting games, extreme weathers, or traditional festivals, the influence of traffic accidents is comparable

Table 2.3 Part of recent transportation applications on social media

Application	Author and year	Social media
Travel information retrieval	Xiang and Gretzel [5]	Google search, Twitter
	Ueno et al. [6]	Twitter
	Evans-Cowley and Griffin [7]	Twitter
	Lee et al. [8]	Twitter
	Lin et al. [9]	Twitter
	Sadri et al. [10]	Twitter
	Chen [11–13]	Brightkite
Social event detection, Traffic incident detection	Sakaki et al. [14]	Twitter
	Krstajic et al. [15]	Twitter
	Schulz et al. [16]	Twitter
	Zhang et al. [17]	Twitter
	Sadri et al. [18]	Twitter
Disaster relief	Gao et al. [19]	Ushahidi
	Ukkusuri et al. [20]	Twitter
	Sadri et al. [21]	Twitter
	Kryvasheyeu et al. [22, 23]	Twitter
	Sadri et al. [24, 25]	Twitter
Traveler behavior, travel pattern	Hasan et al. [26]	Twitter, Foursquare
	Hasan et al. [27, 28]	Twitter
	Wall et al. [29]	Facebook
Transportation planning Transportation policy-making, commercial service, transit management	Camay et al. [30]	Facebook, Twitter, Flickr, etc.
	Stambaugh [31]	Facebook, Twitter, YouTube, etc.
	Gelernter et al. [32]	Twitter
	Pender et al. [33]	Facebook, Twitter
	Chan and Schofer [34]	Twitter
	Gelernter et al. [32]	Twitter
	Ni et al. [35]	Twitter

to a "midget" [9, 47]. From our observations, accident-related tweets are in small quantity. What's more, most of them are confined to a small area and limited to a relatively short time interval, and some researchers call them small-scale events [16]. Second, the texts online are inherently complex and unstructured. The common methods for detecting the traffic-related events include Support Vector Machine [16, 46], natural language processing [48, 49], etc., which explore the semantic features in key words. However, as the context of a tweet is limited to 140 words and the tweet contents try to be concise, key word detection is sometimes not sufficient for accurate automatic language processing.

The common techniques to overcome these challenges are to disintegrate the tweet sentences into a bag of words first, then select the proper features to construct a structured database, and finally employs a classification model to properly identify the tweets.

Table 2.4 Demographics of Twitter users

Attribute	Source					
Education	Pew Research Center	Less than high school	High school	Trade or some college	Bachelor's degree	Graduate school
		6%	16%	39%	21%	18%
Age	Statista.com	18–24	25–34	35–44	45–54	55+
		18.20%	22.20%	20%	16.70%	22.90%
Gender	Statista.com	Male	Female			
		48%	52%			
Income	United States Census Bureau & Pew Research Center	0–30 K	30–50 K	50–75 k	75 K+	
		23.15%	8.35%	13.51%	54.98%	

Table 2.5 Tweet samples describing the general traffic information, general traffic incident, and road accident, respectively

General information	*"I am waiting at the silver line, exciting"* *"Always hate the signals ahead of the hip-hop, making me sick"*
General incident	*"standstill for 1 hour, there must be accidents in front"* *"this is typical NOVA traffic, what a bad day"*
Traffic accident	*"major accident next to the sunoco near the parkway a car got flipped over"* *"the worst car accident possible just happened in front of me"*

2.2.1.1   Step 1 Database Construction

The raw tweet data need to be preprocessed to constitute a database that can be used for further analysis. Usually, the accident-related tweets should contain one or more key words such as "accident" or "crash." However, there has been no consensus on such a vocabulary of the accident-related words. Thus, we turn to the traditional news media and collect about 100 articles of news that discuss the traffic accident. In all these articles, we select the words that appear the most frequently. The frequency of a word is the times that a specific word appears in these articles. These accident-related words include *"police, accident, traffic, crash, road, car, vehicle, highway, driver, county, injured, pm, state, injuries, scene, hospital, according, people, died, near, patrol, morning, happened, dead, taken, just, driving, department, involved, vehicles, south, passenger, hit, truck, north, monday, left, lanes, lane, killed, struck, southbound, area, closed, investigation."*. By applying a filter based on these key words, we can obtain a large quantity of potential tweets. On these tweets, we looped the following procedures to filter out the non-accident-related tweets and obtain the related words:

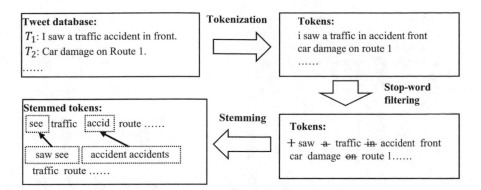

Fig. 2.1 Steps of token filtering and stemming

- Randomly select tweets from the filtered tweets.
- Manually label them whether they are accident-related.
- Extract the most frequent words in accident-related tweets.
- Filter the tweets based on the frequent words.

We finally extracted more than 900 labeled tweets from the New York Metropolitan Area and North Virginia and also paired over 1800 non-accident-related tweets with the accident-related ones to constitute a database. Each tweet post needs to be preprocessed following the procedure in Fig. 2.1. The stop-word list we used refers to Ranks-NL [50]. After the procedure, the tweets are disintegrated into more than 20,000 tokens.

2.2.1.2   Step 2 Feature Selection

The phi coefficient [51] can be employed to calculate the correlation between the tokens and the manual label shown in Eq. (2.1). Those tokens whose $|\phi|$ is higher than 0.1 are selected and part of them are shown in Fig. 2.2.

$$\phi = \frac{n_{11}n_{00} - n_{10}n_{01}}{\sqrt{n_{1*}n_{0*}n_{*0}n_{*1}}} \tag{2.1}$$

where all notations are defined in the following table, x and y denote the manual label and label feature, respectively.

	$y = 1$	$y = 0$	Total
$x = 1$	n_{11}	n_{10}	n_{1*}
$x = 0$	n_{01}	n_{00}	n_{0*}
Total	n_{*1}	n_{*0}	n

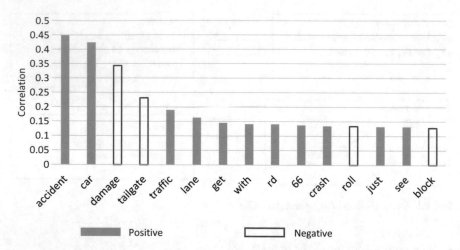

Fig. 2.2 Correlations between the manual label and the individual stemmed tokens. To make it easy to read, we write the basic form of the word instead of the stemmed token

Following this rule, 27 tokens are selected. Some of the tokens may be accounted by the geographic uniqueness such as "66," "95," and "495," indicating the route number where traffic accidents occur. Some may be directly topic-related words including "traffic," "car," "accident," etc. Other words such as "damage" and "tailgate" are too general in our daily life and thus provide negative indicators in describing the traffic accident.

Features from individual tokens are sometimes not sufficient because these emphasize solely the correlations between the label and tokens and may overlook the associations within the tokens. One can further extract paired token features using the Apriori algorithm [52, 53] which finds the regularities in large-scale binary data by two major probabilities: support and confidence.

$$\text{supp}\left(t_j\right) = \frac{\text{size of}\left(\{T_i, \quad t_j \subseteq T_i\}\right)}{\text{size of}\left(\{T_i\}\right)} \tag{2.2}$$

$$\text{supp}\left(t_{j_1} \cap t_{j_2} \cap \ldots \cdots \cap t_{j_m}\right) = \frac{\text{size of}\left(\{T_i, t_{j_1} \cap t_{j_2} \cap \ldots \cdots \cap t_{j_m} \subseteq T_i\}\right)}{\text{size of}\left(\{T_i\}\right)} \tag{2.3}$$

When support is equal to 0.01 and confidence is equal to 0.1, we can find 36 token pairs listed in Table 2.6.

2.2.1.3 Step 3 Classification Model

An effective language modeling method is necessary to extract the useful accident-related information from tweets. In this example, we employ Deep Neural Network

Table 2.6 Paired tokens selected by the Apriori algorithm

accident	block	car	get	i	crash	car
damage	brain	i	get	just	car	get
car	with	accident	get	just	accident	get
tailgate	game	tailgate	i	i	car	get
car	saw	damage	just	accident	car	get
accident	lane	car	just	i	accident	get
accident	traffic	i	just	i	just	car
damage	do	accident	just	just	accident	car
i	do	damage	i	i	just	accident
i	roll	car	i	i	accident	car
car	crash	accident	car			
i	crash	accident	i			
just	get	accident	lane			

(DNN), one of the simplest forms of Deep Learning, in training and classifying the accident-related tweets. DBN consists of densely connected layers, and each layer has a few neurons that represent the activation function. There exist links between neurons from different levels of layers while there is no link between neurons in the same level of layers. Connections between neurons in the same layer may not be practical and have scalability issues. Thus, the DBN in this chapter is also known as Restricted Boltzmann Machine (RBM). The neural functions and basic structure are shown in Fig. 2.3. In a 2-layer neural network, the input is the token features while the output is the manual label. The 2-layer neural network resembles the Artificial Neural Network, similar to that of the Support Vector Machine or logistic regression in which the output value is the direct computation of the input features. In a multilayer neural network, however, the input features are first converted into hidden features as shown in Fig. 2.3b and then the hidden layers finally calculate the corresponding output.

In Fig. 2.3a, the number of categories that a neuron can output is equal to the number of neurons in the upper level. The relationship between the input and the output can be written as Eq. (2.4):

$$b_i = g\left(f_{j,i}\left(W_{j,i}, a_j\right)\right) = g\left(\sum_j W_{j,i} a_j\right) \qquad (2.4)$$

where a_j is a vector for the jth input token feature; b_i is the output; $W_{j,i}$ is the conversion parameter matrix to be estimated; $g()$ is the activation function and can be changed in different levels of neurons. Here we employ the softmax function shown in Eq. (2.5):

$$g\left(f_{j,i}\right) = \frac{e^{f_{j,i}}}{\sum_i e^{f_{j,i}}} \qquad (2.5)$$

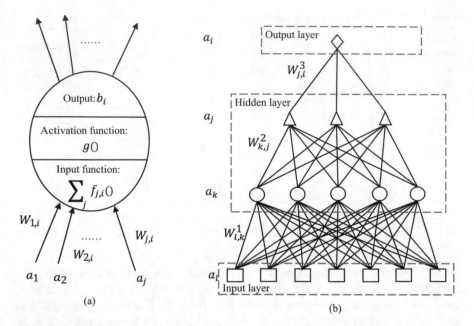

Fig. 2.3 (**a**) Example of a single neuron. $W_{j,i}$ denotes the transition matrix. (**b**) Structures of 4-layer neural networks

For the output layer, with an initial estimate of $W_{j,i}$, one can calculate the square error between the true label and the estimated label as shown in Eq. (2.6). Thus, the regression problem can be converted into an optimization problem in which one can find the best $W_{j,i}$ to minimize the square error δ^2, or diminish the changes of δ^2 until $\nabla(\delta^2)$ smaller than a threshold value by a gradient method or the Newton–Raphson method [54].

$$\delta^2 = \frac{1}{2} \sum_i (y_i - b_i)^2 \tag{2.6}$$

The gradient method is an iterative approach that each cycle finds a descent direction and update the $W_{j,i}$ by a step size. The descent direction can be calculated as

$$\frac{\partial \left(\delta^2 \right)}{\partial W_{j,i}} = \sum_i \left(y_i - g_i \left(\sum_j W_{j,i} a_j \right) \right) \bullet \frac{\partial}{\partial W_{j,i}} \left(y_i - g_i \left(\sum_j W_{j,i} a_j \right) \right) = \sum_i -e_i g_i' a_j \right) \tag{2.7}$$

where $e_i = y_i - g \left(\sum_j W_{j,i} a_j \right)$ is the difference between the predicted label and true label; the $W_{j,i}$ can be updated according to the perceptron learning rule [55]:

$$W_{j,i}{}^{t+1} = W_{j,i}{}^{t} + \alpha \frac{\partial\left(\delta^2\right)}{\partial W_{j,i}} = W_{j,i}{}^{t} + \alpha \sum_i e_i g_i' a_j \tag{2.8}$$

where α is a scale parameter to be decided and t indicates the iteration cycle. W between other layers can be updated in the same way. For the hidden layer, we can update the corresponding $W_{l,k}$ or $W_{k,j}$ by the error from the output layer. The algorithm employed is called back-propagation. The process of back-propagation can be generalized as follows: When the features are placed in the input layer, the effects of the input features are propagated forward through the layer structure, layer by layer until reaching the output layer. By comparing with the true label, using the error function in Eq. (2.6), the error values are then propagated backward, updating the conversion matrix as shown in Eq. (2.9).

$$\frac{\partial\left(\delta^2\right)}{\partial W_{k,j}} = \sum_i - \left(y_i - g_i \left(\sum_j W_{j,i} a_j \right) \right) \bullet \frac{\partial}{\partial W_{k,j}} \left(g_i \left(\sum_j W_{j,i} a_j \right) \right)$$

$$= \sum_i -e_i g_i' W_{j,i} g_j' a_k \tag{2.9}$$

2.3 Result

By setting the confidence to be 0.8 for feature selection of paired tokens, there will be 17 paired token features and totally 16 individual tokens in the paired token features. We can finally obtain good regression results combining these paired token features and the individual token features. From Fig. 2.4, when we set the ϕ to be 0.2, there are only 4 qualified individual token features and the accuracy can be around 0.8. Higher ϕ may result into a simpler model but relatively less accuracy, while lower ϕ improves the performance but may cause overfitting. Thus, one may seek a balanced model in the future applications. It is worth mentioning that the number of neurons in the second and third layers is 10 and 5, respectively. Our examinations show that by changing the number of neurons in two layers, the computing time may be influenced but the accuracy is almost unchanged.

Using Deep Learning methods, one can automatically classify the tweets to acquire a very high accuracy. The accident detection based on social media, especially tweets, is meaningful to the traffic management and potentially complements the current accident detection. The comparisons between the accident-related tweets with both the traffic accident log and loop-detector data indicate some merits of tweets: From this example, it is found that nearly 30% of the accident-related tweets can be located by the accident log and more than 80% of them can be related to abnormal traffic data [56].

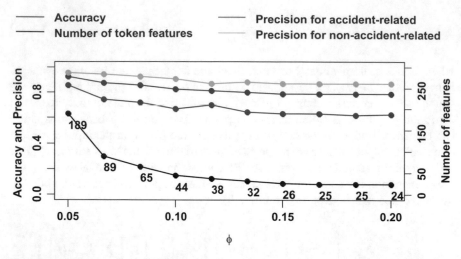

Fig. 2.4 Regression results of DBN with selected individual and paired tokens under different thresholds of correlation coefficient ϕ

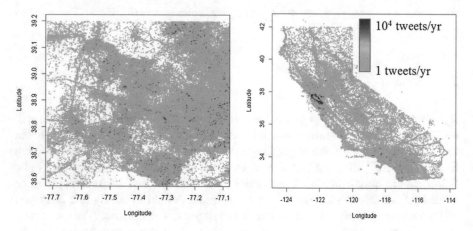

Fig. 2.5 Coverage of tweets in North Virginia and the State of California in 2014. Each point represents a 100 m^2 area

2.3.1 Human Mobility Exploration

Some of the famous social media tools or websites are so popular that they have high coverage over the geographic span. Figure 2.5 shows the coverage of Twitter in North Virginia and the State of California in the whole year of 2014. Thousands, even millions, of tweets assemble in almost all densely populated areas.

For human mobility studies based on Twitter, the displacements generated by Twitter have several advantages: First, in recent years, more and more users prefer to tweet on mobile devices instead of on PCs [57]. The geo-location and time

information of Twitter and its popularity among people make possible the retrieval
of wide-range displacement information from the broad masses of people in a timely
manner. Second, the resolution of the traditional data is not as high as that of Twitter.
For instance, the banknotes [58] are usually used to trace the interstate or intercity
travel and the corresponding time gaps are usually more than days. Popular data
sources from cell phones [59] which can access a more frequent mobility data but
the geo-precision is in kilometers because the average service area of each mobile
tower is approximately 3 km^2 [60].

One can take each Twitter user u as a reliable data source and his tweets as a
spatial-temporal text database: $U = [U_1, U_2, \ldots, U_i, \ldots]$. U_i represents ith text
entry in the database: $U_i = [D_i, L_i, T_i]$ where D_i is the tweet content; L_i is the
location information of latitude and longitude, together with the label the Census
Designated Place (CDP); T_i is the time. The displacement between two time-
sequential locations can be calculated as $|L_i, L_j|$, where $|\ |$ calculates the map
distance. Given an U_i, its corresponding sequential displacements over a certain
time period can be calculated as shown in Eq. (2.10):

$$S_i^W = \left[|L_i, L_j| : W^E \geq T_j - T_i \geq W^S \right] \tag{2.10}$$

where W^E and W^S are the ending and starting time of the time window W. L_j and T_j
refer to all the locations and timestamps of the Twitter user. Thus, S_i^W can be taken
as the domain of all displacements which start from L_i and end after a certain time
interval between W^E and W^S; and all the tweet contents that are posted during these
displacements can be notated in Eq. (2.11):

$$D_i^W = \left[D_j : W^E \geq T_j - T_i \geq W^S \right] \tag{2.11}$$

For each U_i, its longest displacement during W can be notated as the featured
displacement of U_i: $S_i^W = \max \left(S_i^W \right)$ and the domain of S_i^W is notated as
$S^W = \text{dom} \left(S_i^W \right)$ which is the collection of all featured displacements of the
Twitter user. The corresponding U_i of a featured displacement can be notated as
$U_i^W = \left[D_i^W, L_i^W, T_i^W \right]$ where L_i^W and T_i^W are the featured (longest) L_i and T_i
within time window W; and the domain of U_i^W is noted as $U^W = \text{dom} \left(U_i^W \right)$. Here
we define the mobility patterns of a Twitter user as $M^W = \text{domain} \left(M_i^W \right)$ where

$$M_i^W = \left[S_i^W, L_i, L_i^W, T_i, T_i^W \right] \tag{2.12}$$

From Eq. (2.12), the mobility pattern of each user can be characterized by trip
displacements, times, and locations. According to our methods, there may exist
more than one U_i corresponding to the same U_i^C as the Twitter users may tweet
more than one time during a time window W. Thus, for a set of $[M_i^W]$ that has the
same L_i^W, we only keep one M_i^W whose $T_i^W - T_i$ is the largest. This filtering process
can reserve all destinations of the trip displacements instead of all origins for further

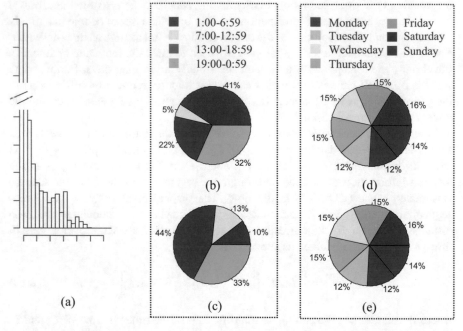

Fig. 2.6 (**a**) Histogram of the displacements over 1-h time window of all Twitter users in this study; the time distribution of the displacements (**b**) lower than 100 m and (**c**) higher than 10,000 m; the day-of-week distribution of the displacements (**d**) lower than 100 m and (**e**) higher than 10,000 m

study. After this process, there will not be any two mobility patterns: $\left[M_i^W, M_j^W \right]$ in which $T_i \geq T_j$ and $T_i^W \leq T_j^W$. The mobility patterns can cover all the possible destinations of the Twitter users together with its corresponding displacement over a certain time window.

Our empirical examinations even find the consecutive hourly displacements for more than 400 Twitter users. The distribution of displacements of all Twitter users in our study may capture not only uniformed decay characteristics of displacement frequency but also a population-based heterogeneity as shown in Fig. 2.6.

We can see that an ever-dominant portion of displacements lower than 100 m which accounts for about 74% of all displacement records. Besides, unlike the cell phones or bank notes, Twitter is sometimes an entertainment tool more than a necessary communication media, and the Twitter records with no arrangements may be much more common. Consequently, Twitter can capture an extremely larger portion of small displacements as compared to other data sources. One can see that nearly half of them are made after midnight, and a very small portion be done in the morning. This observation indicates that people are less likely to make a trip after midnight or in the morning. As compared, for the long displacements shown in Fig. 2.6c, the portions for both morning and midnight shrink to 23%. It is worth

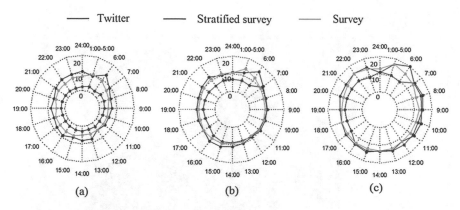

Fig. 2.7 Comparisons of the hourly median of Twitter displacement/trip length higher than (**a**) 1 mile, (**b**) 4 miles, and (**c**) 7 mile. The scaleplate of the radiation plot is (0, 10, and 20 miles) [61]

mentioning that displacements over 2- and 3-h time windows also have the same features as shown in Fig. 2.6. Also, there is almost no day of week features in Twitter displacements.

Figure 2.7 compares the time-of-day displacement features between Twitter and travel survey. The extracted displacements follow additional strict rules introduced in [61] for comparison:

• We only keep Twitter displacements higher than 1/9 mile as the smallest trip length recorded by the survey is 1/9 mile (178 m).
• We only focus on the travel within a metropolitan area and do not include the intercity or interstate trips.

As Twitter may represent a certain group of people with designated distributions of age, gender, education, and household income, we collect user demographics from the open data sources from comScore [36], from Pew Research Center [37], as well as population demographics of Fairfax County in Northern Virginia [62]. We conduct a stratified sampling on the records in household travel survey. Both the original survey and stratified survey are put into comparison in Fig. 2.7. The features of extracted Twitter displacements resemble that of the household travel survey to some extent. One can see that Twitter displacements that contain lots of small displacements underestimate the actual trip length in general. If we only focus on the displacements longer than 4 and 7 miles, one can see a great resemblance between the stratified survey and Twitter.

Due to the large number of Twitter users, the short-distance displacements capture unprecedented details of human travel even though each of them is short and without detailed trajectory information. If we aggregate all these short displacements, they will unveil ever-elaborate depictions of the Northern Virginia networks as shown in Fig. 2.8a. By comparing with the authentic road information, we can prove the validity of these short-distance travels as shown in Fig. 2.8b, c.

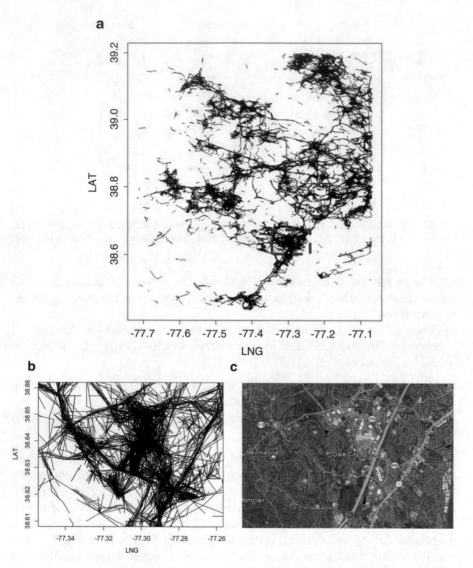

Fig. 2.8 (**a**) Geographic distribution of short-distance displacements in Northern Virginia (the square noted area is Dale City); (**b**) Short-distance displacements in Dale City area; (**c**) road networks in Dale City area

2.3.2   Trip Purpose and Travel Demand Forecasting

Inferring individuals' activity and trip purposes is critical for transportation planning and travel behavior analysis. With the thriving growth of geo-coded land use data and increasing popularity of mobile devices using satellite-based positioning system, a shift happens in the related studies from an active individual or household travel survey to a more passive integral approach. For instance, the mobile phone records obtained from the mobile towers can reveal the human mobility pattern and trajectory features [60], interurban trip patterns [63], etc. Besides, the Global Positioning System (GPS) is also a well-accepted data source in extracting the information of trips or even tours: Wendy et al. [64] combined the GIS data, GPS logs, and the individual characteristics to interpret and validate the travel patterns; Anastasia et al. [65] designed TraceAnnotator system that processes multiday GPS traces semiautomatically to impute transportation modes, activity episodes, and other facets of activity; an experiment in [66] using the trajectories data recorded by a passive GPS summarizes trip activities as "home," "work," "education," "shopping," and "other."

The tweets posted by the GPS-based smartphone can passively collect location-time data. This data collection is not based on an experimental design and is a better solution than the external GPS devices in the travel-survey studies [67]. Thus, the GPS information and corresponding text information collected by tweets can be possibly used to infer the trip purposes. Figure 2.9 gives an example of a Twitter user's locations from 18:00 p.m. to 19:00 p.m. across the year of 2014. We selected three displacements, and one can see that the corresponding tweets clearly unveil the trip purposes.

The trip purpose information found by social media can be very useful to the researchers. Because traditional land use category is too general to provide detailed useful trip information. Figure 2.10 shows a map including a trip end, the parcel of land use, and all commercial places within it. The category of the land use is "Non-Residential Mixed Use." In comparison, the places within it give more detailed information.

One can see that the tweets can be taken as representatives of social awareness toward the public activities and social communications. One can assume that the frequencies of tweet locations may, to some degree, represent the people's location preferences. The location preferences toward a specific event can, in turn, enable us to forecast the travel demand within a local road network. Our exemplary social events are the newly built 11-mile extension of Sliver Line, a subway line of Washington Metro starting service on July 26, 2014. We extract the tweets from July 26 to July 28, which mentioned the names of the metro line and 5 new stations: "Silverline," "Spring Hill," "Greensboro," "Tysons Corner," "McLean," and "Wiehle-Reston East." Except "Silverline," the rest of the topic words are proprietary not only to the metro stations but also to the commercial facilities around the Silver Line. By using the human mobility extraction method in Sect. 2.2.1, one can easily obtain the corresponding Twitter displacements related to the key words and track the origins of these displacements.

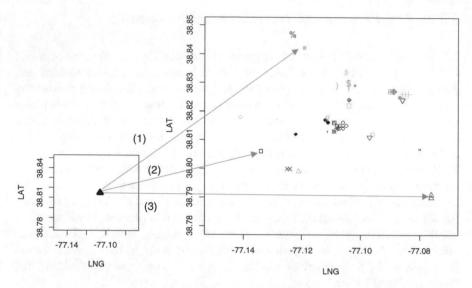

Fig. 2.9 Displacements that originate from (**a**) 18:00 p.m. and end in (**b**) 19:00 p.m. and the corresponding tweets for location: (1) "i hate this haircut"; (2) "ive never seen this movie that they playing on bet"; (3) "xisthatnigga marchmadness this is marchmadness this happening to duke makes this so great"

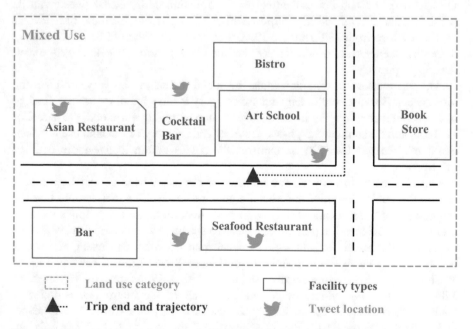

Fig. 2.10 The comparison between land use and tweets in inferring the trip purposes

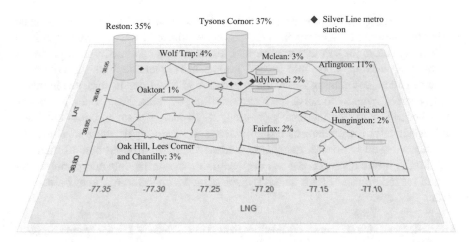

Fig. 2.11 The ratios of CDPs where the "Silver Line" displacements start (origin) from July 26 to July 28

Figure 2.11 shows the census-designated places (CDPs) of the starting locations and traffic is induced from these places by the social event. A ratio is calculated to compare the different responses of CDPs toward the opening of Silver Line. The figure gives a clear picture of geographic distributions of induced transportation demands. In this example, the time window is set to be 1 h, and most of the CDPs are consequently within 10 miles of the metro stations. One can even employ longer time windows and larger geographic scales in future studies.

One can see that the mobility patterns extracted from Twitter enlighten OD estimations for transportation demand analysis. The results come from the social reflections of Twitter instead of the costly, large-scale travel survey. It will be of greater practical significance in future studies with the penetration of Twitter together with other social media tools.

2.4   Discussions on Future Improvements and Applications

This chapter starts with the discussion of social media exploration and uses statistics to introduce its data features and recent studies. In the following section, we employ several examples to detail how the social media can contribute to the current transportation applications and studies in mainly three topics: human mobility, event detection, as well as the trip purpose and travel demand forecasting. The examples prove the promises of Twitter for transportation researchers mainly because it has two categories of information: (1) Twitter provides a vast amount of location and time information which is both of high resolution and high accuracy. (2) Twitter contributes large-quantity text information which is a reflection of the social awareness from massive crowds; as a consequence, monitoring the social media data may deliver useful traffic event information.

There are also limitations on the studies. For instance, when used for demand estimation purposes, social media data should be adjusted for overrepresentation of such system users [68]. Future transportation studies on social media shall gradually narrow the research scope into several important respects such as the automatic place detection, commuting behaviors in urban road networks. In sum, social media study opens a window to solve the transportation problems. Further studies can be more accurate with the increasing coverage of tweets as a social communication tool. Relatively, researchers can also increase the geographic span to study the intercity, or even interstate, travels, as well as the trip purposes behind them. In addition, in future attempts, we can even combine the traffic data and social media data to see the impact of long-distance displacements on the travel time, traffic flow throughput, etc.

Acknowledgement This study was partially supported by the National Science Foundation award CMMI-1637604.

References

1. R. Buettner, Getting a job via career-oriented social networking sites: the weakness of ties, in *2016 49th Hawaii International Conference on System Sciences (HICSS)*, 2016, pp. 2156–2165
2. DMR, 2016, http://expandedramblings.com/.
3. T. Aichner, F. Jacob, Measuring the degree of corporate social media use. Int. J. Mark. Res. **57**, 257–275 (2015)
4. Z. Zhang, *Fusing Social Media and Traditional Traffic Data for Advanced Traveler Information and Travel Behavior Analysis* (State University of New York at Buffalo, Buffalo, 2017)
5. Z. Xiang, U. Gretzel, Role of social media in online travel information search. Tour. Manag. **31**, 179–188 (2010)
6. K.S.S.N.K. Ueno, K. Cho. Feasibility study on detection of transportation information exploiting twitter as a sensor, 2012
7. J. Evans-Cowley, G. Griffin, Microparticipation with social media for community engagement in transportation planning. Transp. Res. Rec.: J. Transp. Res. Board **2307**, 90–98 (2012)
8. J.H. Lee, S. Gao, K. Janowicz, K.G. Goulias, Can Twitter data be used to validate travel demand models?, in *IATBR 2015-WINDSOR*, 2015
9. L. Lin, M. Ni, Q. He, J. Gao, A.W. Sadek, T. I. T. I. Director, Modeling the impacts of inclement weather on freeway traffic speed: an exploratory study utilizing social media data, in *Transportation Research Board 94th Annual Meeting*, 2015
10. A.M. Sadri, S. Hasan, S.V. Ukkusuri, Joint Inference of User Community and Interest Patterns in Social Interaction Networks, *arXiv preprint arXiv:1704.01706*, 2017
11. Y. Chen, H.S. Mahmassani, Use of social networking data to explore activity and destination choice behavior in two metropolitan areas. Transp. Res. Rec.: J. Transp. Res. Board **2566**, 71–82 (2016)
12. Y. Chen, A. Talebpour, H.S. Mahmassani, Friends don't let friends drive on bad routes: modeling the impact of social networks on drivers' route choice behavior, in *Transportation Research Board 94th Annual Meeting*, 2015
13. Y. Chen, H.S. Mahmassani, Exploring activity and destination choice behavior in two metropolitan areas using social networking data, in *Transportation Research Board 95th Annual Meeting*, 2016

14. T. Sakaki, M. Okazaki, Y. Matsuo, Earthquake shakes twitter users: real-time event detection by social sensors, in *Proceedings of the 19th international conference on World wide web*, 2010, pp. 851–860
15. M. Krstajic, C. Rohrdantz, M. Hund, A. Weiler, Getting there first: Real-time detection of real-world incidents on twitter, 2012
16. A. Schulz, P. Ristoski, H. Paulheim, I See a Car Crash: Real-Time Detection of Small Scale Incidents in Microblogs, in *The Semantic Web: ESWC 2013 Satellite Events*, (Springer, Berlin, 2013), pp. 22–33
17. S. Zhang, J. Tang, H. Wang, Y. Wang, Enhancing traffic incident detection by using spatial point pattern analysis on social media. Transp. Res. Rec.: J. Transp. Res. Board **2528**, 69–77 (2015)
18. A.M. Sadri, S. Hasan, S.V. Ukkusuri, J.E.S. Lopez, Analyzing Social Interaction Networks from Twitter for Planned Special Events, *arXiv preprint arXiv:1704.02489*, 2017
19. H. Gao, G. Barbier, R. Goolsby, D. Zeng, Harnessing the crowdsourcing power of social media for disaster relief, DTIC Document 2011
20. S. Ukkusuri, X. Zhan, A. Sadri, Q. Ye, Use of social media data to explore crisis informatics: study of 2013 Oklahoma Tornado. Transp. Res. Rec.: J. Transp. Res. Board **2459**, 110–118 (2014)
21. A.M. Sadri, S. Hasan, S.V. Ukkusuri, M. Cebrian, Understanding information spreading in social media during Hurricane Sandy: user activity and network properties, *arXiv preprint arXiv:1706.03019*, 2017
22. Y. Kryvasheyeu, H. Chen, E. Moro, P. Van Hentenryck, M. Cebrian, Performance of social network sensors during Hurricane Sandy. PLoS One **10**, e0117288 (2015)
23. Y. Kryvasheyeu, H. Chen, N. Obradovich, E. Moro, P. Van Hentenryck, J. Fowler, et al., Rapid assessment of disaster damage using social media activity. Sci. Adv. **2**, e1500779 (2016)
24. A.M. Sadri, S.V. Ukkusuri, H. Gladwin, The role of social networks and information sources on hurricane evacuation decision making. Nat. Hazards Rev. **18**, 04017005 (2017)
25. A.M. Sadri, S.V. Ukkusuri, H. Gladwin, Modeling joint evacuation decisions in social networks: the case of Hurricane Sandy. J. Choice Modell. **25**, 50–60 (2017)
26. S. Hasan, S.V. Ukkusuri, X. Zhan, Understanding social influence in activity location choice and lifestyle patterns using geolocation data from social media. Front. ICT **3**(10) (2016)
27. S. Hasan, S.V. Ukkusuri, Urban activity pattern classification using topic models from online geo-location data. Transp. Res. Pt. C **44**, 363–381 (2014)
28. S. Hasan, X. Zhan, S.V. Ukkusuri, Understanding urban human activity and mobility patterns using large-scale location-based data from online social media, in *Proceedings of the 2nd ACM SIGKDD international workshop on urban computing*, 2013, p. 6
29. T. Wall, G. Macfarlane, K. Watkins, Exploring the use of egocentric online social network data to characterize individual air travel behavior. Transp. Res. Rec.: J. Transp. Res. Board **2400**, 78–86 (2013)
30. S. Camay, L. Brown, M. Makoid, Role of social media in environmental review process of national environmental policy act. Transp. Res. Rec.: J. Transp. Res. Board **2307**, 99–107 (2012)
31. C. Stambaugh, Social media and primary commercial service airports. Transp. Res. Rec.: J. Transp. Res. Board **2325**, 76–86 (2013)
32. J. Gelernter, S. Balaji, An algorithm for local geoparsing of microtext. GeoInformatica **17**, 635–667 (2013)
33. B. Pender, G. Currie, A. Delbosc, N. Shiwakoti, International study of current and potential social media applications in unplanned passenger rail disruptions. Transp. Res. Rec.: J. Transp. Res. Board **2419**, 118–127 (2014)
34. R. Chan, J. Schofer, Role of social media in communicating transit disruptions. Transp. Res. Rec.: J. Transp. Res. Board **2415**, 145–151 (2014)
35. M. Ni, Q. He, J. Gao, Forecasting the subway passenger flow under event occurrences with social media. IEEE Trans. Intell. Transp. Eng. **18**, 1623–1632 (2017)
36. L. Adam, L. Andrew, 2016 U.S. cross-platform future in focus, 2016

37. M. Duggan, J. Brenner, *The Demographics of Social Media Users, 2012*, vol 14 (Pew Research Center's Internet & American Life Project, Washington, DC, 2013)
38. Statista, *Distribution of Twitter users in the United States as of December 2016, by age group*, 2017, https://www.statista.com.
39. Statista, Number of active Twitter users in the United States from 2010 to 2014, by gender (in millions), 2016
40. H. Purohit, A. Hampton, S. Bhatt, V.L. Shalin, A. Sheth, J. Flach, An information filtering and management model for twitter traffic to assist crises response coordination, *Special Issue on Crisis Informatics and Collaboration*, 2013
41. E. Aramaki, S. Maskawa, M. Morita, Twitter catches the flu: detecting influenza epidemics using Twitter, in *Proceedings of the conference on empirical methods in natural language processing*, 2011, pp. 1568–1576
42. C. Shirky, The political power of social media. Foreign Aff. **90**, 28–41 (2011)
43. E. Mai, R. Hranac, Twitter interactions as a data source for transportation incidents, in *Proc. Transportation Research Board 92nd Ann. Meeting*, 2013
44. A. Gal-Tzur, S.M. Grant-Muller, T. Kuflik, E. Minkov, S. Nocera, I. Shoor, The potential of social media in delivering transport policy goals. Transp. Policy **32**, 115–123 (2014)
45. Y. Gu, Z.S. Qian, F. Chen, From twitter to detector: real-time traffic incident detection using social media data. Transp. Res. Pt. C **67**, 321–342 (2016)
46. E. D'Andrea, P. Ducange, B. Lazzerini, F. Marcelloni, Real-time detection of traffic from twitter stream analysis. Intell. Transp. Syst. IEEE Trans. **16**, 2269–2283 (2015)
47. Z. Zhang, M. Ni, Q. He, J. Gao, J. Gou, X. Li, An exploratory study on the correlation between twitter concentration and traffic surge 2. Transp. Res. Rec.: J. Transp. Res. Board **35**, 36 (2016)
48. N. Wanichayapong, W. Pruthipunyaskul, W. Pattara-Atikom, P. Chaovalit, Social-based traffic information extraction and classification, in *ITS Telecommunications (ITST), 2011 11th International Conference on*, 2011, pp. 107–112
49. R. Li, K. H. Lei, R. Khadiwala, and K. C.-C. Chang, Tedas: a twitter-based event detection and analysis system, in *Data engineering (ICDE), 2012 IEEE 28th international conference on*, 2012, pp. 1273–1276
50. Ranks-NL, Default English stopwords list, 2015, http://www.ranks.nl/stopwords.
51. H. Cramér, *Mathematical Methods of Statistics*, vol 9 (Princeton University Press, Princeton, 1999)
52. R. Agrawal, R. Srikant, Fast algorithms for mining association rules. in *Proc. 20th int. conf. very large data bases, VLDB*, 1994, pp. 487–499
53. M. Hahsler, B. Grün, K. Hornik, Introduction to arules–mining association rules and frequent item sets. in *SIGKDD Explor*, 2007
54. T.J. Ypma, Historical development of the Newton–Raphson method. SIAM Rev. **37**, 531–551 (1995)
55. Y. Freund, R.E. Schapire, Large margin classification using the perceptron algorithm. Mach. Learn. **37**, 277–296 (1999)
56. Z. Zhang, Q. He, J. Gao, M. Ni, A. Deep Learning, Approach for detecting traffic accidents from social media data. Transp. Res. Pt. C **86**, 580–596 (2016)
57. E. Protalinski, More Twitter users chose to tweet from a mobile device rather than a PC in 2012, study says, 2012, http://thenextweb.com/twitter/2013/04/12/more-twitter-users-chose-to-tweet-from-a-mobile-device-rather-than-a-pc-in-2012-study-says/#gref.
58. D. Brockmann, L. Hufnagel, T. Geisel, The scaling laws of human travel. Nature **439**, 462–465 (2006)
59. C. Song, Z. Qu, N. Blumm, A.-L. Barabási, Limits of predictability in human mobility. Science **327**, 1018–1021 (2010)
60. M.C. Gonzalez, C.A. Hidalgo, A.-L. Barabasi, Understanding individual human mobility patterns. Nature **453**, 779–782 (2008)
61. Z. Zhang, Q. He, S. Zhu, Potentials of using social media to infer the longitudinal travel behavior: a sequential model-based clustering method. Transp. Res. Pt. C **85**, 396–414 (2016)

62. K. Fatima, P. Anne, H. Cahill M.L. Erik, B. Khamthakone, Demographic Reports 2015, County of Fairfax, Virginia. in *Countywide Service Integration and Planning Management (CSIPM), Economic, Demographic and Statistical Research January, 2016*, 2016
63. C. Kang, X. Ma, D. Tong, Y. Liu, Intra-urban human mobility patterns: an urban morphology perspective. Phys. A: Stat. Mech. Appl. **391**, 1702–1717 (2012)
64. W. Bohte, K. Maat, Deriving and validating trip purposes and travel modes for multi-day GPS-based travel surveys: a large-scale application in the Netherlands. Transp. Res. Pt. C **17**, 285–297 (2009)
65. A. Moiseeva, J. Jessurun, H. Timmermans, Semiautomatic imputation of activity travel diaries: use of global positioning system traces, prompted recall, and context-sensitive learning algorithms. Transp. Res. Rec.: J. Transp. Res. Board **2183**, 60–68 (2010)
66. L. Shen, P.R. Stopher, A process for trip purpose imputation from global positioning system data. Transp. Res. Pt. C **36**, 261–267 (2013)
67. L. Stenneth, O. Wolfson, P.S. Yu, B. Xu, Transportation mode detection using mobile phones and GIS information. in *Proceedings of the 19th ACM SIGSPATIAL International Conference on Advances in Geographic Information Systems*, 2011, pp. 54–63
68. T.H. Rashidi, A. Abbasi, M. Maghrebi, S. Hasan, T.S. Waller, Exploring the capacity of social media data for modelling travel behaviour: opportunities and challenges. Transp. Res. Pt. C **75**, 197–211 (2017)

Chapter 3
Ground Transportation Big Data Analytics and Third Party Validation: Solutions for a New Era of Regulation and Private Sector Innovation

Matthew W. Daus

3.1 Part I: History of Taxi Data

3.1.1 Overview of the New York City T-PEP Program

In March 2004, the New York City Taxi and Limousine Commission ("TLC") mandated that specific technology-based improvements known as the Taxicab Passenger Enhancement Program ("T-PEP") be implemented in all medallion taxicabs.[1] The costs of the T-PEP system were essentially offset by a 26% fare increase, the largest increase in the history of the TLC.[2]

The author is the longest serving former Commissioner/Chair and General Counsel of the New York City Taxi and Limousine Commission, and currently serves as: Distinguished Lecturer at The United States Department of Transportation's University Transportation Research Center (UTRC Region 2—New York, New Jersey & Puerto Rico) at The City College of New York, of the City University of New York (CUNY); pro bono President of the International Association of Transportation Regulators (IATR); and Partner & Chair of the Transportation Practice Group at the law firm of Windels Marx Lane & Mittendorf, LLP.

[1] See http://www.nyc.gov/html/tlc/html/industry/taxicab_serv_enh_archive.shtml.

[2] See Taxi & Limousine Commission 2004 Annual Report to the City of New York, http://www.nyc.gov/html/tlc/downloads/pdf/2004_annual_report.pdf.

M. W. Daus (✉)
City College of New York, New York, NY, USA
e-mail: mdaus@windelsmarx.com

© Springer International Publishing AG, part of Springer Nature 2019   47
S. V. Ukkusuri, C. Yang (eds.), *Transportation Analytics in the Era of Big Data*,
Complex Networks and Dynamic Systems 4,
https://doi.org/10.1007/978-3-319-75862-6_3

The yellow medallion[3] taxicabs are the only for-hire vehicles allowed by law to pick up passengers by street hail in New York City.[4] Since 2008, the TLC has required all yellow taxicabs to be equipped with T-PEP. Medallion owners and drivers are strictly prohibited from tampering with or removing the T-PEP system. Anyone found guilty of tampering with the permanent modifications or shutting off the system during business will be subject to fines and suspension. Drivers are permitted to work with a broken system for up to 48 h as long as they have reported the problem and are awaiting repair. The TLC has set forth the requirement that at least 90% of all system repairs must be completed within 6 h.[5]

Despite mandating the use of T-PEP, the TLC is not responsible for the installation, operation, and maintenance of the system. Instead, these functions are outsourced to external vendors who meet requirements set forth by the TLC. Since 2004, several vendors were authorized, but as of now only Creative Mobile Technologies and Verifone remain as technological providers of T-PEP. In 2013, the TLC passed new rules with updated requirements for T-PEP providers. Flywheel Software[6] is the latest addition to the list of authorized providers and has been live since September 2016.[7]

T-PEP is essentially comprised of four components: (1) Driver Information Monitor ("DIM"); (2) Passenger Information Monitor ("PIM"); (3) Credit/Debit Card Payment System; and (4) Trip Sheet Automation.[8] In addition, a wheelchair accessible taxi must be equipped with a system for accepting dispatches through TLC's Accessible Dispatch program. The hardware of these components are typically disparate and hard-mounted onto the vehicles with extensive wiring. In the near future, however, these components might be integrated, with less physical devices, into what is known as the Alternative Technology Solution ("ATS").[9]

The DIM is used to relay messages to drivers only when the vehicle is moving very slowly or has stopped, to prevent them from being distracted. The TLC sends short alphanumeric messages to drivers who can then respond by pushing single-button pre-programmed responses. This is particularly useful in the event of a

---

[3] A medallion refers to a small metal plate attached to the hood of a taxi that confers the driver the right to pick up any passengers within New York City.

[4] NYC Admin. Code §19-504; New York State adopted the HAIL Act which allowed certain livery vehicles to do street hails outside of the Manhattan business district and airports (Chapter 602 of the Laws of 2011, as amended by Chapter 9 of the Laws of 2012 ("HAIL Act")).

[5] See http://www.nyc.gov/html/tlc/html/passenger/passenger_creditcard.shtml.

[6] In early April 2017, Flywheel was acquired by its competitor Cabconnect through an undisclosed deal. See Ken Yeung, Cabconnect acquires Flywheel in bid to create on-demand taxi platform, VentureBeat, Apr. 7, 2017, https://venturebeat.com/2017/04/07/cabconnect-acquires-flywheel-in-bid-to-create-on-demand-taxi-platform/.

[7] Erica Jackson, Flywheel App Comes to New York City, NYC Biz News, Oct. 11, 2016, http://nycbiznews.journalism.cuny.edu/2016/10/flywheel-app-comes-to-new-york-city/.

[8] See http://www.nyc.gov/html/tlc/html/passenger/taxicab_serv_enh.shtml.

[9] The ATS will quite likely exclude the PIM in light of constant complaints from drivers and passengers regarding its lack of responsiveness and redundancy.

citywide emergency, traffic congestion, or road closures. It may also be utilized to direct drivers to return lost property or to locations with fare opportunities.[10]

The PIM, also known as Taxi TV, is a flat screen monitor installed in the back seat of the cab that displays a map of the vehicle's current location in addition to advertisements, news, weather forecast, and the TLC's Public Service Announcements ("PSAs"). The map on the screen only shows the route traveled and not the projected path, nor does it provide any information regarding traffic conditions. Passengers who dislike the repetitive content or noise can turn off the system, but it will be reactivated at the end of the ride to assist in payment.[11]

The card payment system, also located in the rear of the cab, provides passengers with a high level of security as far as the transaction is concerned. Once a card is swiped, it takes about 5 s for the payment to be processed. There is no minimum fee imposed, but customers are required to sign the receipt when the fare is $25 and above. Tips can be entered on the touch screen. All card payment systems are certified by the Payment Card Industry ("PCI").[12]

The trip sheet automation, which is the focus of the next section, allows the collection and submission of trip data using an Automatic Vehicle Locator ("AVL"). It could rightly be identified as the transportation technology that ushered the taxicabs of New York City into the age of big data. Back in the old days, taxicab drivers were required to record trip-logs of every fare manually (i.e., pen and paper). They were also required by the TLC to maintain these trip sheets for at least three years. Not only was this tedious work, the data recorded were also prone to human error. With AVL, however, data is captured automatically as soon as the driver turns on the taximeter. The electronic data collected includes date, time, trip distance, itemized fares,[13] and pick-up and drop-off locations.[14] Only the number of passengers and payment type are entered manually. Once the driver turns off the taximeter at the end of the trip, the data is transmitted to the TLC's data servers.

3.1.2 T-PEP Data Collection and Output

The TLC updates the trip data on their Research & Statistics webpage every six months with two months' lag. For example, trip data from January to June 2017 would only be available publicly by August 2017. Figure 3.1 shows a snapshot

---

[10] *See* http://www.nyc.gov/html/tlc/html/industry/taxicab_serv_enh_archive.shtml.

[11] *Id.*

[12] The PCI is an information security standard that protects cardholder data and maintains a secure network for organizations that handle branded credit cards.

[13] This can be broken down into time-and-distance fare, MTA tax, tips, tolls, extras and surcharges. It does not include any cash tips.

[14] Since July 2016, the TLC only provides taxi zones instead of precise latitude/longitude coordinates for pick-ups and drop-offs. The taxi zones include 265 unique neighborhood areas.

Metric	Description
Trips per Day	Average number of trips recorded each day
Farebox per Day	Total amount, across all vehicles, collected from all fares, surcharges, taxes, and tolls. Note: this amount does not include amounts from credit card tips
Unique Drivers	The total unique number of hack drivers who recorded a trip each month
Unique Medallions	The total unique number of medallion taxis and standby vehicles* who recorded at least one trip in the month
Medallions Per Day	The average unique number of medallion taxis and standby vehicles* who recorded at least one trip in a day
Avg Days Medallions on Road	The average number of days each vehicle spent on the road per month
Avg Hours Per Day Per Medallion	The average number of hours in which a vehicle recorded a trip
Avg Days Drivers on Road	The average number of days each driver recorded a trip
Avg Hours Per Day Per Driver	The average number of hours each
Avg Minutes Per Trip	Average trip time from meter-on to meter-off
Percent Trips Paid with Credit Card	Number of trips where passenger paid by credit card out of the total number of trips

*Stand-by vehicles are back-up vehicles permitted for use by fleets when medallion taxis are out of service.

Fig. 3.1 Snapshot of trip sheet data. Source: http://www.nyc.gov/html/tlc/html/about/trip_record_data.shtml

of yellow taxicab trips in December 2016. Since the dataset contains details of over ten million trips within that month, the trip sheet can only be viewed in its entirety through an open source database and not a typical spreadsheet. There are roughly 1.3 billion trip records in the current database.[15] The TLC also publishes spreadsheets containing "metrics[,] including average daily trips and fares collected, active vehicles and drivers, and credit card usage in yellow taxis tabulated from yellow taxi trip data collected through the Taxi Passenger Enhancement Program (TPEP)."[16]

In addition to the trip records mentioned above, the TLC also provides monthly indicators such as number of trips, fares collected, active vehicles and drivers, and credit card usage on their website. In contrast to the trip sheet data, this is easily viewable using any spreadsheet program such as Excel, since there are limited numbers of rows.[17] Based on this dataset, trend charts of key variables that are indicative of the medallion market can be plotted (see Figs. 3.2 and 3.3). Last but not least, the TLC has thus far published two "Factbooks," one in 2014 and another in 2016, which provide readers with infographics of trends of trips and fare.

---

[15] *See* Chicago Taxi Data, https://github.com/toddwschneider/chicago-taxi-data.

[16] These are available at http://home2.nyc.gov/html/tlc/html/about/statistics.shtml.

[17] *See Id.*

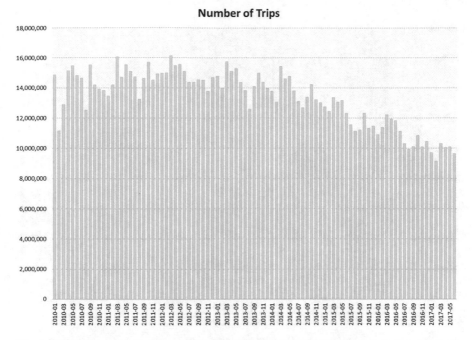

Fig. 3.2 Number of trips from January 2010 to January 2018. Source: Author, based on data from TLC

Thanks to the successful implementation of the T-PEP in medallion taxicabs, the TLC has expanded the data collection requirements to include Street Hail Liveries ("SHL"), resulting in the Livery Passenger Enhancement Program ("L-PEP"), and For-Hire Vehicles ("FHVs"). For the former, trip records and monthly indicators are available since 2013. The metrics captured are more or less similar to those of T-PEP. For the latter, trip data is available since 2015, with significantly less metrics—only dispatching base number, date, time, and location of pick-ups are captured. Nonetheless, as a late adopter of the program, FHVs have been able to take advantage of newer technologies and leapfrog to the ATS, requiring fewer modifications and hardware—in other words, less costs.

3.1.3   T-PEP and Privacy Protection

On December 13, 2012, the New York City Taxi and Limousine Commission ("TLC") promulgated rules for the Authorization of T-PEP Providers.[18] Chapter 76 of the TLC Rules and Regulations sets forth information security standards that

---

[18]*See* http://rules.cityofnewyork.us/tags/tpep.

Farebox Revenue

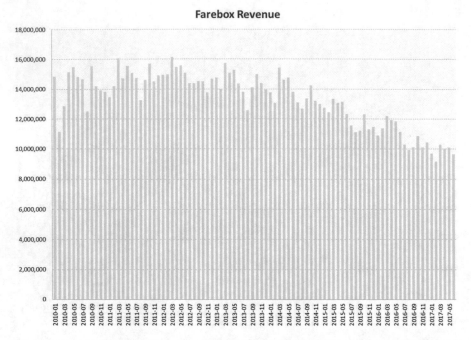

Fig. 3.3 Monthly farebox revenue from January 2010 to January 2018. Source: Author, based on data from TLC. Note: Revenue includes all fares, surcharges, taxes and tolls; but not credit card and cash tips

T-PEP systems had to meet in order to be approved by the TLC for sale, lease, or use in taxicabs. Under these regulations, T-PEP Providers were required (1) to establish policies for information security, authentication, remote access, anti-virus security, application development security, digital media re-use and disposal, encryption, passwords, user responsibilities, and vulnerability management; (2) to comply with copyrights and develop appropriate controls and procedures to protect the Database Management Systems[19]; (3) to limit access to T-PEP Data, by providing safeguards such as firewalls and fraud prevention; (4) to maintain the confidentiality of personal information; and (5) to develop controls for network management and procedures for security incident management.[20]

The TLC repealed Chapter 76 in 2016 when it promulgated new rules requiring all TLC licensees and authorized service or equipment providers that collect a passenger's personal information or geolocation information—including T-PEP

---

[19]A Database Management System was defined under the rules as "a software package with computer programs that control the creation, maintenance and use of a database." *See* Chapter 76 of the TLC Rules and Regulations, Sec. 76-02 (repealed).

[20]Chapter 76 of the TLC Rules and Regulations, *"Information Security Rules for Taxicab Technology Service Providers,"* repealed in 2016.

Providers—to comply and file with the TLC an Information Security and Use of Personal Information Policy.[21] The new rules require such policies to include, at a minimum, the following information:[22]

- A statement of internal access policies relating to passengers' and drivers' personal information[23] for employees, contractors, and third party access, if applicable;
- A statement that, except to the extent necessary to provide credit, debit, and prepaid card services and services for any application that provides for electronic payment, personal information will only be collected and used with such passenger's affirmative express consent and that such personal information will not be used, shared, or disclosed, except for lawful purposes;
- Procedures for notifying the TLC and affected parties of any breach of the security of the system, pursuant to New York law;[24]
- A statement that any credit, debit, or prepaid card information collected by the T-PEP Provider or a credit, debit, or prepaid card services provider is processed by the T-PEP Provider or such provider in compliance with applicable payment card industry standards; and
- A statement of the T-PEP Provider's policies regarding the use of passenger geolocation information, which must include, at a minimum, a prohibition on the use, monitoring, or disclosure of trip information, including the date, time, pick-up location, drop-off location, and real-time vehicle location and any retained vehicle location records, without such passenger's affirmative express consent.

In addition, the collection, transmission, and maintenance of data by T-PEP Providers must comply with applicable PCI Standards[25], as well as New York City Department of Information Technology and Telecommunications Citywide Information Security Policy for Service Providers and Encryption Standards ("DOITT Standards").[26]

---

[21] *See* http://www.nyc.gov/html/tlc/downloads/pdf/proposed_rules_fhv_bills_package.pdf; *See also* R.C.N.Y. Title 35, Chapter 75, Sec. 75-23, *Business Requirements—Use of Personal Information and Certain Location-Based Data* (effective 08/06/2016).

[22] R.C.N.Y. Title 35, Chapter 75, Sec. 75-05 (b)(2)(i).

[23] *See* R.C.N.Y. Title 35, Chapter 75, Sec. 75-03 (ee). (Any information that can specifically identify an individual, such as name, address, social security number, unmasked or non-truncated credit, debit, or prepaid card numbers, together with any other information that relates to an individual who has been so identified, and any other information that is otherwise subject to privacy or confidentiality laws and associated rules and regulations. The display or disclosure of only the last four digits of a credit, debit, or prepaid card number is not Personal Information. The name of a Taxicab Driver and the Driver's Commission license number is not Personal Information.)

[24] *See* New York General Business Law, Section 899-aa.

[25] The Payment Card Industry Data Security Standards issued by the Payment Card Industry Security Standards Council may change from time to time. *See* www.pcisecuritystandards.org.

[26] *See* R.C.N.Y. Title 35, Chapter 75, Sec. 75-25(f).

The TLC is entitled to only a limited amount of data which includes taxicab pick-up and drop-off data, as well as certain GPS location information. The TLC does not typically obtain, and is generally prevented from reviewing, breadcrumb data, or the GPS pings of the taxicab and its location throughout the route in between pick-up and passenger drop-off. This is precisely the type of information, the tracking of a passenger trip, that Uber was alleged to have been monitoring as part of its "God View" tool.

The TLC typically obtains very important T-PEP data on the number of rides, the taxi fare information, and other general information that include "blips or dots on a screen." They have no particular identity of passengers or individual taxicab drivers or medallions, unless requested for a specific legitimate regulatory purpose as part of a TLC or government investigation. Off-duty locations of taxicabs are completely off limits to the TLC as a privacy safeguard. The TLC agreed with the NYCLU that this safeguard was embodied in the T-PEP vendor agreements and the TLC rules.

The TLC collects general ridership data to achieve various objectives, such as to verify that taxicabs are servicing all neighborhoods in the city. Also, the data determines the actual earnings of taxicab drivers and medallion owners, which is used to make sound fact-based decisions in determining fare increases. This eliminates the prior guesswork involved in manual trip sheet surveys and other primitive regulatory methods. The GPS systems have been instrumental in the return of lost passenger property, alerting drivers without fares to business opportunities in underserved areas, and using "breadcrumb" data to ping taxis throughout on-duty trips to estimate vehicle speed and assess the viability of traffic policies, such as NYC's pedestrian plaza initiatives.[27] The TLC will only receive further breadcrumb data from the T-PEP system if it is specifically requested for a targeted and disclosed purpose (i.e., lost property; stolen cab, etc.). Further, the TLC will only release more detailed data to law enforcement if served with a subpoena.

Despite the use of such information for these and many other legitimate government objectives, drivers, industry, and civil liberties groups have raised objections and commenced lawsuits. In 2007, as the TLC just started to install T-PEP units in NYC taxicabs, taxi driver groups, and other union leaders standing in solidarity, called for a taxi strike alleging the GPS aspect invaded their privacy rights. All of those efforts have failed, however, and the law recognizes that privacy rights in public taxicabs are minimal. When the drivers were unsuccessful in pulling off a strike, they then turned to the courts to sue the TLC, claiming privacy rights violations of the fourth Amendment of the U.S. Constitution, and lost decisively in *Alexandre v. NYC TLC.*[28]

---

[27]Michael M. Grynbaum, *New York Traffic Experiment Gets Permanent Run*, The New York Times, Feb. 11, 2010, http://www.nytimes.com/2010/02/12/nyregion/12broadway.html.

[28]*Alexandre v. NYC TLC,* No. 07 Civ. 8175 (RMB), 2007 U.S. Dist. LEXIS 73642 (S.D.N.Y. Sept. 28, 2007).

In *Alexandre v. NYC TLC*, the Court examined whether the fourth and fifth Amendments of the U.S. Constitution prohibiting government from conducting unlawful searches and takings of private property forbids the TLC from mandating the installation of T-PEP. The roots of the case law stretch back to a decision involving police officers attaching a tracking beeper to a chloroform container following an automobile's movement across state lines in *U.S. v. Knotts*.[29] In this 1983 criminal case, the U.S. Supreme Court held there was "no reasonable expectation of privacy in [an] automobile on public thoroughfares."[30] In *Alexandre v. NYC TLC*, the Court found the TLC's contracts with its vendors limited the release of information about the location of a taxicab while it is off-duty.[31] Applying the fourth Amendment case law, the Court held there was "no legitimate expectation of privacy" when information was readily available for public scrutiny.[32] The Court reasoned, based on *Knotts*, that there is no reasonable expectation of privacy in a motor vehicle traveling on a public roadway, and applied an "intermediate level of scrutiny.[33] The Court concluded that TLC's substantial interest in requiring GPS outweighed privacy rights in that the government was promoting taxi customer service, taxicab ridership, and passenger and driver safety.[34]

The 5th Amendment's "takings' clause" provides that private property shall not be taken for public use without "just compensation."[35] There are two categories of regulatory actions that generally will be deemed *per se* takings for fifth Amendment purposes: (1) a physical taking where the government requires an owner to suffer a permanent physical invasion of his/her property; (2) regulations that completely deprive an owner of "all economically beneficial us[e]" of his/her property.[36] Outside these two relatively narrow categories of *per se* takings, regulatory takings challenges are governed by the standards set forth in *Penn Central Transp. Co. v. New York City*.[37] The Court in *Penn Central* identified "several factors that have particular significance" in evaluating regulatory takings claims.[38] Primary among those factors are "[t]he economic impact of the regulation on the claimant and, particularly, the extent to which the regulation has interfered with distinct investment-backed expectations."[39] In addition, the "character of the governmental action"—for instance whether it amounts to a physical invasion or instead merely affects property interests—may be relevant in discerning whether

---

[29] 460 U.S. 276 (1983).

[30] *Id.* at 281.

[31] *Alexandre*, 2007 U.S. Dist. LEXIS 73642 at 32.

[32] *Id.*

[33] *Id.* at 33-34.

[34] *Alexandre*, 2007 U.S. Dist. LEXIS 73642 at 34.

[35] U.S. Const. Amend. V.

[36] *Lingle v. Chevron U.S.A, Inc.*, 544 U.S. 528, 538 (2005).

[37] *Lingle*, at 538-539, citing *Penn Central Transp. Co. v. New York City*, 438 U. S. 104 (1978).

[38] *Ibid.*

[39] *Ibid.*

a taking has occurred.[40] The *Penn Central* factors have served as the principal guidelines for resolving regulatory takings claims that do not fall within the first two categories.

In response to the Plaintiffs' fifth Amendment argument, the court in *Alexandre* ruled that medallion owners who choose to engage in a "publicly regulated business" surrender their rights to unfettered discretion as to how to conduct same.[41] The Court cited a federal lawsuit commenced in the Eastern District of Pennsylvania, *MCQ's Enterprises v. PPA*, where Plaintiff Yellow Cab argued that being required to install the City's coordinated dispatch system in its cabs without compensation "for the economic injury and loss that will result from the taking of Yellow Cab's customers and intellectual property" constituted an unconstitutional taking.[42] The Pennsylvania Federal Court denied a preliminary injunction as PPA rules were promulgated "for the public good...to promote hospitality and tourism...."[43] Similarly, the Court in *Alexandre* ruled that the NYC TLC rules in issue were enacted to "protect the public interest."[44]

The state of the law on privacy rights is settled in some ways, but untested on other issues. Clearly, regulatory bodies stand on firm legal privacy grounds as compared to criminal law enforcement and employee monitoring. Government regulators need to clearly define the use of any data derived from GPS tracking systems or electronic trip sheets by both specifying the public purposes and interests protected, and restricting the use of such data to those purposes—via agency regulations, contracts, and/or some other tangible way. While the existing case law generally supports the rights of regulators to use electronic trip sheet data, the courts have not yet specifically tested other scenarios, including the use of data obtained while taxicabs are "off duty" and the use of such data for criminal law enforcement investigations.

3.2   Part II: The Advent of the TNC Movement and TNC Data

3.2.1   Overview of Transportation Network Companies

As mobile technology continues to improve, more and more ride-hailing apps are available via smartphones. Five of the most prominent app-based car services in New York City are Uber, Lyft, Gett, Juno,[45] and Via. The rise of these apps is due

---

[40] *Ibid.*

[41] *Alexandre*, 2007 U.S. Dist. LEXIS 73642 at 27.

[42] *McQ's Enters. v. Phila. Parking Auth.*, 2007 U.S. Dist. LEXIS 2130, at 12.

[43] *Id.*

[44] *Alexandre,* 2007 U.S. Dist. LEXIS 73642, at 29.

[45] At the end of April 2017, Juno was acquired by Gett for $200 million. *See* Brian Solomon, *Gett Buys Juno For $200 Million, Uniting Would-Be Uber Competitors*, Forbes,

in part to the ease and conveniences of prompt service, and the ever-increasing use of smartphones for transactions of all types. With 40 million monthly active riders[46] and current worth of $69 billion, Uber is undoubtedly the most aggressive and fastest growing TNC of them all.[47] Lyft comes in second, with a current valuation of $7.5 billion after a recent funding round of $600 million.[48]

Compared to reserving or prearranging car services or taxicabs, the e-hailing process is convenient, and a user only needs to download an app, run it, set a pick-up location, make a booking and proceed with payment. Things are, however, slightly more complicated behind the scenes. In Uber's case, after a user opens the app and authenticates their identity, the app sends a ping with the user's geolocation to their server every 5 s. The server then responds with a list of all the available car types at the user's location. For each car type, the nearest eight cars, expected waiting time, and surge multiplier are provided.[49]

Within 5 years, these Transportation Network Companies ("TNCs") have devoured a big piece of the car service pie, resulting in a slump in the medallion market. Since 2014, the prices of medallions in NYC have fallen significantly and a growing number of medallions have been foreclosed by lenders. Many drivers have complained that there are too many car service providers and not enough passengers for them to make a living. Some have stressed the need for more enforcement because of illegal street hails by TNCs. Others have called for the TLC to impose a cap on TNCs. There is a general consensus that the TLC should hold all sectors to the same standards regarding accessibility, licensing, and vehicle requirements, etc.[50]

There is no question that the competition and disruption created by TNCs, whether legal, ethical or policy-challenged, has set the stage for a paradigm shift for transportation policy planners on a much broader basis. Decades of transportation planning and policy, which sought to minimize and deter personal motor vehicle ("PMV") usage, is now facing a collision course with hundreds of thousands of additional vehicles that continue to be added to congested roads[51]—many operated

---

Apr. 26, 2017, https://www.forbes.com/sites/briansolomon/2017/04/26/gett-buys-juno-for-200-million-uniting-would-be-uber-competitors/#764d63f96089.

[46]Matthew Lynley, *Travis Kalanick says Uber has 40 million monthly active riders*, TechCrunch, Oct. 19, 2016, https://techcrunch.com/2016/10/19/travis-kalanick-says-uber-has-40-million-monthly-active-riders/.

[47]Brad Stone, *The $99 Billion Idea: How Uber and Airbnb Won*, Bloomberg Businessweek, Jan. 26, 2017, https://www.bloomberg.com/features/2017-uber-airbnb-99-billion-idea/.

[48]Heather Somerville, *Lyft lands $600 million in fresh funding; company valued at $7.5 billion*, Reuters, Apr. 11, 2017, http://www.reuters.com/article/us-lyft-funding-idUSKBN17D2I8.

[49]*See* https://www.ftc.gov/system/files/documents/public_comments/2015/09/00011-97592.pdf.

[50]*See* Transcript of April 6, 2017 Taxi & Limousine Commission Board Meeting, http://www.nyc.gov/html/tlc/downloads/pdf/transcript_04_06_2017.pdf.

[51]For the first time since 1990–2014, ridership growth in taxi/FHV has outpaced those of public transit, particularly in 2016. Between 2013 and 2016, TNCs added 600 million vehicle miles traveled, exacerbating congestion in NYC. *See* Bruce Schaller, *Unsustainable? The Growth of*

by less experienced and part-time TNC drivers who do not need to go through rigorous biometric criminal background checks. TNC drivers may be encouraged to work longer hours to achieve the same prior economic benefits, possibly contributing to greater driver fatigue and unsafe working conditions. Also, TNC vehicles are encouraged by so-called "surge pricing" to work during peak demand times—or rush hour—when traffic and environmental conditions may be at their worst.[52]

Surge pricing refers to TNCs increasing their prices in certain areas, or at specific times, in response to local demand. Surge pricing has resulted in nightmares for many consumers who unknowingly agree to pay exorbitant prices for relatively short rides, and then only notice the steep charges until after the ride is complete. This occurs during peak demand times, with the greatest surges often following large events and holiday celebrations. For example, every year on New Year's Day a host of disgruntled consumers share their stories of excessive surge price charges from the night before. Customer receipts show numerous examples in which the "surge" increased the rate to 9.9 times the normal fare, and what would have normally cost a rider $20.71, resulted in a $205.03 charge for the roughly 20 min trip.[53]

In theory, surge pricing takes place when demand for service exceeds the number of available vehicles. TNCs argue that the higher fares incentivize drivers to provide trips when there are more ride requests than drivers looking for fares by encouraging drivers to be available in areas where they typically would not have been otherwise. Predictably, fares that surge to multiple times the average price can have the effect of pricing out certain population segments, resulting in drivers choosing not to operate in certain areas altogether, a practice known as redlining.[54] In other words, drivers may refuse to operate in communities where there is less of an opportunity to earn large fares, thus discouraging drivers from providing services in what have traditionally been underserved areas. Because TNCs strictly control their data—and much of the data they release to the public portrays them in a positive light—it is difficult to definitively determine the net effects of surge pricing on the wider transportation industry, its consumers and stakeholders.

Another area of concern is the lack of oversight regulation as to the calculation of distance and time by TNCs. A smartphone app may not meet the requirements of a taximeter, which is required for taxicabs to calculate fares based on the distance travelled and the time elapsed. For instance, apps are not "wired" into the vehicle

---

App-Based Ride Services and Traffic, Travel and the Future of New York City, Feb. 27, 2017, http://schallerconsult.com/rideservices/unsustainable.pdf.

[52] Dan Kedmey, *This Is How Uber's 'Surge Pricing' Works*, Time, Dec. 15, 2014, http://time.com/3633469/uber-surge-pricing/.

[53] Stephanie McNeal, *People Woke Up and Realized They Spent Hundreds of Dollars On Uber For New Year's Eve*, BuzzFeed, Jan. 1, 2016, https://www.buzzfeed.com/stephaniemcneal/uber-hangover?utm_term=.ehx3W62jqM#.woOg8a4yKN.

[54] "Redlining" refers to the formal or informal practice of establishing geographical borders where service will not be offered.

transmission but, instead, rely on the GPS to calculate the fare. The most significant problem with the use of GPS is accuracy, which is needed to calculate the fare. Additionally, taximeters are calibrated, tested, and sealed by a regulatory authority and require periodic inspections. However, there is no such regulation of GPS in this environment and the method by which a smartphone calculates fares. Because of the lack of weights and measures conformity, consumer protection concerns are raised that TNCs may be charging consumer fares in excess of applicable regulatory limits. Further, some apps dispatch for-hire vehicles. In most jurisdictions, for-hire vehicles must charge fares based on a prearranged basis or in accordance with a filed fare schedule; however, some apps charge passengers like a taxicab, based on distance and mileage (and demand).

The essence of TNC laws revolves around a "we can do it faster and better than government" attitude, which, in terms of efficiency, may be correct given their resources. However, there is an ulterior motive, as no app-based dispatch model ever works without having an adequate supply of drivers. It is simply too costly and difficult to entice and subsidize the transfer of professionally licensed black car and taxicab drivers to TNCs (although this was done successfully at a very high cost by Uber in New York City). TNCs claim that college students and part-time workers would be discouraged by the process of purchasing insurance, completing physical paperwork, leaving their homes or computers and undertaking a simple 5 min fingerprint check. While there is some truth to the convenience factor, the motive of TNCs is to attract more drivers by expanding the pie, making it easier to recruit drivers while managing and assuming the risks of some potentially unsafe or inexperienced drivers who slip through the cracks and cause harm to others. The self-regulation model allows the TNCs to control the information pertaining to public incidents such as sexual assaults and crashes, and discourages further media coverage as such information—which is of public interest—has been labeled as "proprietary," and neither within the government's control nor subject to public disclosure laws that keep such stories and criticism alive. The self-regulation model is an effort to control information, make licensing shortcuts, and to facilitate a market takeover.

It may be possible to engage in modified self-regulation of transportation companies, as is done with trucking and limousine companies engaged in interstate commerce by the U.S. Department of Transportation's Federal Motor Carrier Safety Administration ("FMCSA"). The FMCSA requires that interstate truckers and drivers obtain medical exams,[55] and not work more than a certain number of hours during a time period for public safety reasons.[56] FMCSA-licensed carriers must collect information ensured by Federal auditor compliance.[57] As such, the key ingredients for the success of a self-regulation system are auditing resources, and

---

[55] See https://www.fmcsa.dot.gov/faq/Medical-Requirements.

[56] See https://www.fmcsa.dot.gov/regulations/hours-of-service.

[57] See https://www.fmcsa.dot.gov/international-programs/certification-safety-auditors-safety-investigators-and-drivervehicle.

significant penalties and fines to serve as an appropriate and effective deterrent. Without unbridled access to TNC data to audit real-time performance and compliance at all levels, including the ability to impose significant fines, this model is doomed to fail. In general, governments, and not private parties, should be the ones who regulate; but if TNCs are allowed to engage in self-regulation due to the lack of government resources, they should be required to pay for the enforcement resources and turn over their data to facilitate auditing and compliance.

3.2.2   A Data-Driven Business Model

In order to operate, TNCs collect, retain, and process massive amounts of data with respect to their users, including detailed information on passengers.[58] This information often includes a passenger's name, contact information, payment information, device location, device manufacturer and model, mobile operating system, pick-up location, destination, trip history, contact information for those with whom customers wish to share information, and information about how customers interact with the TNC's interfaces (e.g., browser types and IP addresses).[59]

This data may be more valuable than the transportation services and supports a large portion of the alleged multi-billion dollar valuation of companies such as Uber and Lyft. Of all the issues to surface during the TNC debate so far, whether criminal background checks, insurance, accessibility, unfair competition, etc., nothing could be quite as damning or damaging economically to the new breed of data hungry TNCs than government passing regulations limiting the collection or use of such data. TNCs fundamentally rely on big data for intelligent decision making, from geolocation to fare estimation to surge pricing.[60] Due to the increasing amounts of data collected, and the varied, still forming legal protections of consumer information in the new sharing economy, the purposes for which TNCs collect, store, use, and share consumer data are a matter of public interest.

3.2.3   TNCs' Reluctance to Disclose Data

TNCs have repeatedly refused to hand over their data to regulators, claiming that their business model and technology are proprietary.

---

[58]Hogan Lovells: Review and Assessment of Uber's Privacy Program, p. 3, accessible at https://newsroom.uber.com/wp-content/uploads/2015/01/Full-Report-Review-and-Assessment-of-Ubers-Privacy-Program-01.30.15.pdf.

[59]*Id.*

[60]*How Uber uses data science to reinvent transportation?*, DeZyre, Aug. 4, 2016, https://www.dezyre.com/article/how-uber-uses-data-science-to-reinvent-transportation/290.

In November 2014, the TLC passed rules requiring, among other things, that for-hire vehicle ("FHV") bases submit trip records to the TLC, similar to data requested of medallion taxicabs. At the public hearing on the rules, representatives from both Uber and Lyft testified in opposition to the proposed rules. Uber testified that the collection of data created privacy concerns. Although Uber claimed that these new rules would jeopardize trade secrets and that they were "unconstitutional,"[61] Uber's own privacy policy[62] allows for the sharing of user information, including location data, in response to legal demands. As a result of Uber's refusal to produce the mandated information, the TLC briefly suspended five of Uber's six bases in New York City.[63]

A couple of years later, TNCs again objected to the sharing of data in the context of the adoption of new rules by the TLC designed to reduce the risks of fatigued driving among drivers of for-hire vehicles ("driver fatigue rules").[64] TNCs objected to the data disclosure requirements provided under the new regulations, arguing that these violate passengers' privacy.[65]

The TLC published the final version of these rules on February 13, 2017. The new rules—which went into effect 30 days after this publication—set daily and weekly driving limits that will be calculated based on a method that incorporated feedback from the for-hire vehicle ("FHV") industry. The rules prohibit a taxi or FHV driver from transporting passengers for more than 10 h in any 24-h period. This clock is reset, and the 10 h period begins again, if a driver goes at least 8 consecutive hours without transporting passengers. Drivers cannot transport passengers for more than 60 h in a single calendar week (Sunday to Saturday). Not only are drivers liable for violations when they exceed these limits, but FHV bases are prohibited from dispatching drivers who have exceeded either the daily or weekly limit. If a driver is dispatched from multiple bases or operates as both a taxi driver and an FHV driver, the total number of hours the driver operates either a taxi or FHV (from all bases) will be combined to calculate the daily and weekly hours. However, if a driver exceeds the daily or weekly limit only by combined hours driving for different FHV bases, then only the driver is potentially liable for violations—not the FHV base.

In order to enforce these driver fatigue rules, the TLC requires FHV bases to submit drop-off times and locations in order to track drivers hours (in addition to pick-up times and location FHV bases were previously required to submit).

---

[61]Catherine Yang, *For Now, Business as Usual for Uber in NYC Despite 5 Base Suspensions*, The Epoch Times, Jan. 7, 2015, http://m.theepochtimes.com/n3/1183267-for-now-business-as-usual-for-uber-in-nyc-despite-five-base-suspensions/.

[62]*See* https://www.uber.com/legal/usa/privacy.

[63]Rebecca Harshbarger, *Uber bases suspended after refusing to hand over trip records*, Jan. 7, 2015, http://nypost.com/2015/01/07/uber-bases-suspended-after-refusing-to-hand-over-trip-records/.

[64]*See* http://www.nyc.gov/html/dcas/downloads/pdf/cityrecord/cityrecord-02-13-17.pdf.

[65]Brad Gerstman, *Uber's objections to city's driver-fatigue rules ring hollow*, Crain's New York, Jan. 26, 2017, http://www.crainsnewyork.com/article/20170126/OPINION/170129966/ubers-objections-to-citys-driver-fatigue-rules-ring-hollow.

Unlike the privacy concerns that stem from FHV apps that track the whereabouts of customers after they have finished their rides, the TLC data to be collected will simply be neighborhoods where riders are picked up and dropped off. To protect trip details from the public, the TLC will also obscure license plates and medallions. Consequently, one would never be able to ascertain the destination address or other personal information about the passengers in the vehicle. The Commission already has this information for yellow and green taxi fleets, but lacked drop-off data from the black cars and other FHVs like Uber and Lyft.

In January 2015, Uber announced that it would share anonymized trip data with the city of Boston, including the general area in which trips began and ended (based on the city's zip codes), distance traveled, trip duration, and time stamps.[66] The city discovered issues with the data, including locational data that was too broad for any meaningful analysis,[67] and the data was shared in infrequent bulk downloads, which was not helpful for city planners who were trying to analyze traffic patterns on a daily or weekly basis. The city also criticized the agreement between Uber and the city which restricted the number of agencies that could access the data.[68]

In January 2017, Uber released a new online platform known as "Movement" which provides detailed traffic patterns and travel times. This tool is currently only available for city planners and local governments, but will be made available to the public in the upcoming months.[69] It will be useful for data analysts who seek to learn more about traffic patterns and transportation networks in different cities. Nonetheless, given Uber's shaky track record privacy with user privacy, skeptics are concerned that the anonymized and aggregated individual user data could be reversed-engineered for malevolent purposes such as stalking, identity theft, and financial fraud. Ultimately, it is argued that data anonymization should not be left entirely to Uber but to independent, third-party organizations that represent the privacy interests of users.[70]

---

[66] Adam Vaccaro, *Uber to Hand Over Trip Data to Boston*, Boston.com, Jan. 13, 2015, https://www.boston.com/news/technology/2015/01/13/uber-to-hand-over-trip-data-to-boston.

[67] Adam Vaccaro, *Boston wants better data from Uber, and is taking a roundabout route to try and get it*, Boston.com, June 28, 2016, https://www.boston.com/news/business/2016/06/28/uber-data-boston-wants.

[68] Adam Vaccaro, *Highly touted Boston-Uber partnership has not lived up to hype so far*, Boston.com, June 16, 2016, https://www.boston.com/news/business/2016/06/16/bostons-uber-partnership-has-not-lived-up-to-promise.

[69] Alex Davies, *Uber's Mildly Helpful Data Tool Could Help Cities Fix Streets*, Wired, Jan. 8, 2017, https://www.wired.com/2017/01/uber-movement-traffic-data-tool/.

[70] Julia Franz, *Uber is making ride-booking data publicly available. Is this a privacy Pandora's box?*, Public Radio International, Feb. 4, 2017, https://www.pri.org/stories/2017-01-21/uber-making-ride-booking-data-publicly-available-privacy-pandora-s-box.

3.2.4   TNC's Privacy and Security Issues

3.2.4.1   TNCs' Failure to Protect Consumer Data

In spring 2014, Uber suffered a significant data breach that potentially exposed drivers' names, license numbers, Social Security Numbers as well as bank account and routing numbers.[71] Uber did not discover the breach until September 2014, and only notified the drivers in February 2015.[72] In March 2015, it was reported that thousands of Uber customer logins were available for purchase for as little as $1 each on the anonymous "dark web."[73] Uber users have since allegedly been charged for various fraudulent rides.[74] Uber, however, rejected that these were the source of the stolen logins.[75]

In January 2016, the New York State Attorney General announced a settlement with Uber requiring the car service app to protect riders' personal information.[76] The settlement required Uber to encrypt rider geolocation information and adopt other data security practices (among other things). Uber's privacy and security issues, however, remain unresolved. In October 2016, following a pattern of complaints from Uber customers who have purportedly been charged for trips that they did not take, security professionals called on the ride-sharing company to conduct an investigation to determine if its database had been breached.[77]

The TNC business model also allows for drivers to provide transportation services for multiple companies at the same time. A driver can accept a ride from a

---

[71] Uber Statement Update, posted by Katherine Tassi on June 17, 2016, available at: https://newsroom.uber.com/statement-update/. In March 2015, a former Uber driver based in Portland, Oregon filed a lawsuit against Uber alleging that the company failed to secure and safeguard its drivers' personally identifiable information, including names, drivers licenses numbers and other personal information, and failed to provide timely and adequate notice to Plaintiff and other class members that their private information had been stolen, in violation of California state law (*Antman v. Uber*, Case No. 3:15-cv-01175-JCS (N.D. Ca)). Uber subsequently filed a John Doe lawsuit in an attempt to identify the perpetrator of the breach (*Uber Technologies, Inc. v. John Doe I*, No. C 15-00908 LB (N.D. Cal. March 16, 2015)).

[72] Federal Trade Commission, *In re: Uber and Consumer Privacy*, EPIC Complaint, June 22, 2015, p. 12.

[73] Joseph Cox, *Stolen Uber Customer Accounts Are for Sale on the Dark Web for $1*, Motherboard, March 27, 2015, http://motherboard.vice.com/read/stolen-uber-customer-accounts-are-for-sale-on-the-dark-web-for-1.

[74] Ramzy Alwakeel, *Londoner hit with £3000 cab bill after 'hackers' rack up 142 Uber journeys*, The Evening Standard, March 30, 2015, http://www.standard.co.uk/news/london/londoner-hit-with-3000-taxi-bill-after-hackers-rack-up-142-uber-journeys-10144655.html.

[75] Robert Hackett, *Stolen Uber user logins are for sale on the dark web: only $1 each*, Fortune, March 30, 2015, http://fortune.com/2015/03/30/uber-stolen-account-credentials-alphabay/.

[76] *Uber agrees to enhance user privacy in NY AG settlement*, Jan. 7, 2016, http://pix11.com/2016/01/07/uber-agrees-to-enhance-user-privacy-in-ny-ag-settlement/.

[77] Shanifa Nasser, *Privacy experts call on Uber to investigate after man gets nearly $1000 bogus bill*, CBC News, Oct. 25, 2016, http://www.cbc.ca/news/canada/toronto/privacy-experts-call-on-uber-to-investigate-after-man-gets-nearly-1000-bogus-bill-1.3819640.

passenger through the Uber app for one trip, and then accept a ride through the Lyft app for the next trip, using the same device for both. This adds a new element of concern with regard to data security and who is able to access consumer data.

3.2.4.2 TNCs' Misuse of Customer Data

In 2011, it was reported that Uber was tracking the whereabouts and movements of customers and projecting such information on screens for entertainment at the company's launch parties.[78] A year later, an Uber official revealed in a blogpost how he would supposedly analyze anonymous ridership data in several cities across the U.S. in order to predict customers' overnight sexual liaisons—which Uber referred to as "Rides of Glory."[79]

In November 2014, reports surfaced that an Uber senior executive had allegedly suggested that his company could spend $1 million on "digging up dirt" about unfavorable reporters,[80] and that an Uber official had tracked a reporter's movements without her permission.[81] Similarly, another journalist claimed that she was warned by sources at Uber that executives could be spying on her via her Uber usage.[82] These various news reports drew attention to Uber's "God View" tool that allowed Uber employees to obtain the real-time and historic location data for a ride. In the wake of these privacy scandals, Uber issued a statement insisting that it had a strict policy prohibiting all employees at every level from accessing a rider's or driver's data, with the exception of a "limited set of legitimate business purposes."[83]

Uber can now potentially track the location of all its users, even when they are not using the Uber app. Following an update launched in November 2016, the company is asking customers permission to collect location data from the moment they request a ride until 5 min after their trip ends, including when

---

[78] Kashmir Hill, *'God View': Uber Allegedly Stalked Users For Party-Goers' Viewing Pleasure (Updated)*, Forbes, Oct. 3, 2014, http://www.forbes.com/sites/kashmirhill/2014/10/03/god-view-uber-allegedly-stalked-users-for-party-goers-viewing-pleasure/#9f7f2d23f84e.

[79] Craig Timberg, Nancy Scola, and Andrea Peterson, *Uber Executive Stirs Up Privacy Controversy*, The Washington Post, Nov. 18, 2014, https://www.washingtonpost.com/business/technology/uber-executive-stirs-up-privacy-controversy/2014/11/18/d0607836-6f61-11e4-ad12-3734c461eab6_story.html?utm_term=.b239144d86de.

[80] Ben Smith, *Uber Executive Suggests Digging Up Dirt On Journalists*, BuzzFeed, Nov. 17, 2014, https://www.buzzfeed.com/bensmith/uber-executive-suggests-digging-up-dirt-on-journalists?utm_term=.tf3eA9nYa#.cuBZoLOlx.

[81] Johana Bhuiyan and Charlie Warzel, *"God View": Uber Investigates Its Top New York Executive For Privacy Violations*, BuzzFeed, Nov. 18, 2014, https://www.buzzfccd.com/johanabhuiyan/uber-is-investigating-its-top-new-york-executive-for-privacy?utm_term=.yla05NMnx#.syoVgKDQZ.

[82] Ellen Cushing, *Uber Employees Warned a San Francisco Magazine Writer That Executives Might Snoop on Her*, San Francisco Magazine, Nov. 19, 2014, http://www.modernluxury.com/san-francisco/story/uber-employees-warned-san-francisco-magazine-writer-executives-might-snoop-her.

[83] https://newsroom.uber.com/ubers-data-privacy-policy/.

the app is running in the background of the customer's phone.[84] Uber laid the groundwork for this app update in 2015 when it updated its privacy policy to allow for collection of background location data, which sparked criticism from privacy advocates and led the Electronic Privacy Information Center ("EPIC") to file a complaint with the FTC.[85] Geolocation data, when linked together with the vast array of personal data already collected by TNCs, offers an invaluable insight into an individual's life. In an effort to increase transparency about how it handles location information, Uber announced in April 2017 the implementation of a new privacy settings menu in its app which, among other things, would help users review their location settings, and will make it easier for users to delete their accounts. Uber also vowed to permanently delete users' data 30 days after they have deleted their account.[86]

Uber has repeatedly pointed out that it has clear privacy and security policies against the access to, and use of, personal data outside of legitimate business purposes. These purportedly include: facilitating payment transactions for drivers, monitoring driver and rider accounts for fraudulent activities, and reviewing specific rider or driver accounts in order to troubleshoot bugs. The data-handling practices highlighted above appear inconsistent with Uber's policies. It remains unclear whether Uber has set up reliable data security protections. At the end of November 2014, Uber hired a law firm to conduct an internal data-privacy review, in a move intended to address the public backlash sparked by reports of Uber's controversial use of customer data.[87] The privacy report was released in January 2015. Overall, the review of Uber's privacy practices was positive, and the privacy assessment called for additional actions, including increased employee privacy training, improved transparency regarding the use of customer data and enhanced access controls.[88] Uber stated in early 2015 that it would implement all of these recommendations. It should be noted that this report only focused on the review of

---

[84] Andrew J. Hawkins, *Uber wants to track your location even when you're not using the app*, The Verge, Nov. 30, 2016, http://www.theverge.com/2016/11/30/13763714/uber-location-data-tracking-app-privacy-ios-android.

[85] *In the Matter of Uber Technologies, Inc.*, Federal Trade Commission 152 3054, Complaint, Request for Investigation, Injunction, and Other Relief Submitted by The Electronic Privacy Information Center, https://epic.org/privacy/internet/ftc/uber/Complaint.pdf.

[86] *See* Uber Newsroom, *Your Privacy Settings: All in One Place and Easier to Use* (Apr. 28, 2017), https://newsroom.uber.com/your-privacy-settings-all-in-one-place-and-easier-to-use/; *See also* Kate Conger, *Uber adds privacy info and easy account deletion*, TechCrunch (Apr. 28, 2017), https://techcrunch.com/2017/04/28/uber-adds-privacy-info-and-easy-account-deletion/.

[87] Ellen Rosen, *Uber Hires Hogan Lovells for Review: Business of Law*, Bloomberg, Nov. 21, 2014, https://www.bloomberg.com/news/articles/2014-11-21/uber-hires-hogan-lovells-for-review-business-of-law.

[88] Jedidiah Bracy, *Uber To Implement Privacy Program Recommendations*, International Association of Privacy Professionals, Feb. 2, 2015, https://iapp.org/news/a/uber-to-implement-privacy-program-recommendations/.

the company's written privacy policies, rather than on their actual implementation and enforceability. Some critics have questioned the credibility of this report.[89]

Uber's illegal and predatory behavior is no accident.[90] When some jurisdictions sought to enforce the existing for-hire vehicle regulations against Uber, Uber retaliated by developing a software program specifically designed to circumvent law enforcement operations.[91] As law enforcement officials issued cease and desist letters, impounded vehicles, and issued penalties, Uber developed and employed software and surveillance equipment to identify law enforcement officials to impede their investigations.[92]

Uber is not the only ride-hailing company that has raised concerns about its mishandling of customers' information. In November 2014, a reporter claimed that a Lyft executive had allegedly accessed her trip log information.[93] Following this report, Lyft announced a change in its internal privacy policies to limit employee access to user data by instituting "tiered access controls" that would limit access to user data to a subset of employees and contractors, with access to ride location data restricted to an even smaller subset of people.[94]

3.2.4.3   Customer Data Communication to Third Parties

In February 2014, Miguel Garcia, a ride-hailing app user, brought a putative class action against Lyft and Enterprise Holdings Inc., the former and present owner of the ridesharing application "Zimride."[95] Garcia alleged that each time he used the app, it automatically disclosed his personal information such as his gender, age, zip code, travel plans, link to his Facebook profile and other data to a third party analytics company called Mixpanel. Garcia further contended that Mixpanel would typically use consumers' personal information to compile

---

[89]Evan Schuman, *Uber shows how not to do a privacy report*, Computerworld, Feb. 5, 2015, http://www.computerworld.com/article/2880596/uber-shows-how-not-to-do-a-privacy-report.html.

[90]Michael Horn, *Uber, Disruptive Innovation And Regulated Markets*, Forbes, June 20, 2016, https://www.forbes.com/sites/michaelhorn/2016/06/20/uber-disruptive-innovation-and-regulated-markets/#140ed07237fb.

[91]Mike Isaac, *How Uber Deceives the Authorities Worldwide*, The New York Times, March 3, 2017, https://www.nytimes.com/2017/03/03/technology/uber-greyball-program-evade-authorities.html.

[92]*Id.;* Lori Aratani, *Virginia officials order Uber, Lyft to stop operating in the state*, The Washington Post, June 5, 2014, https://www.washingtonpost.com/news/dr-gridlock/wp/2014/06/05/virginia-officials-order-uber-lyft-to-stop-operating-in-the-state/?utm_term=.64e77d1054e8.

[93]Liz Gannes, *It's Not Just Uber: Tech Companies Snooping on Users Is All Too Common*, Recode, Nov. 20, 2014, http://www.recode.net/2014/11/20/11633100/tech-companies-snooping-on-users-creepy-and-common.

[94]Liz Gannes, *Lyft Limits Employee Access to Data After Re/code Report*, Recode, Nov. 21, 2014, http://www.recode.net/2014/11/21/11633164/lyft-limits-employee-access-to-data-after-recode-report.

[95]*Garcia v. Enter. Holdings, Inc.*, 78F. Supp. 3d 1125 (N.D. Cal., January 23, 2015).

comprehensive profiles of consumers' entire digital lives, which could be then sold as a commodity.[96] Garcia relied upon a specific provision of the California Invasion of Privacy Act, which prohibits any person who, in the course of business, acquires or has access to personal information concerning an individual, "for the purpose of assisting private entities in the establishment or implementation of carpooling or ridesharing programs" from disclosing that information to any other person or using that information for any other purpose without the prior written consent of the individual.

The Court found, however, that this provision did not apply to the case in issue since Enterprise and Lyft were not "assisting private entities" under the law, but the actual creators of the app.[97] The Court also pointed out that Garcia had failed to allege that Lyft, Enterprise or Zimride disclosed his information to Mixpanel for any purpose other than establishing or implementing a rideshare or carpooling program.

Accordingly, the Court granted Lyft and Enterprise's motion to dismiss, and dismissed Garcia's complaint (with permission granted to file an amended complaint). Although Lyft and Enterprise were not found to have violated any privacy law in this case, the latter illustrates how ride-hailing companies may transfer their users' personal data to third party companies, which raises questions as to the security safeguards implemented for such transfers and by the TNCs' contractors themselves.

Communication of individuals' personal information by TNCs to third-party entities may not be limited to the private sector. According to its own transparency report released in April 2016, Uber allegedly received 33 regulatory requests in the second half of 2015 involving trip data for more than 12 million Uber users.[98] Uber was allegedly subpoenaed for rider data 312 times, for driver data 138 times, and for general data 267 times between July 2015 and December 2015.[99] Uber supposedly handed over some data in more than 82% of those instances. The company also purportedly received 90 search warrants, 30 emergency requests, 28 court orders and produced information in about 80% of those cases.

3.2.4.4 Recommendations

So far, ride-hailing companies have pushed back against strong security protections in favor of unbridled growth. However, as TNCs have become data repositories with insight into individuals' personal and work lives, how these companies protect

---

[96] *Id.* at 1130.

[97] *Id.* at 1134.

[98] *See* https://newsroom.uber.com/transparency-report-2015/.

[99] Duncan Macrae, *Uber Driver/Customer Data Sharing Raises Privacy Concerns*, TechWeekEurope, April 13, 2016, http://www.techweekeurope.co.uk/e-regulation/legal/uber-drivercustomer-data-sharing-raises-privacy-concerns-189921#HatvxQXEXXj52KA3.99.

consumer data is important. TNC's potential unlawful tracking of passengers and consumer data collection along with lapses in privacy safeguards are troubling. The Federal Trade Commission ("FTC") and other regulators have become increasingly concerned with the privacy implications of mobile and geolocation data and mobile app data security.[100]

Whether changes are on the way on a national legislative level, it is completely within the power of state and local legislators or government transportation regulators to require, as a condition to the licensure of TNCs, that privacy protections be put into place. To the extent privacy measures are currently in place for technology used in taxicabs and/or limousines, TNCs should be held to the same standard. As jurisdictions enact new TNC legislation, or revisit such legislation, it is incumbent on our lawmakers to ensure that appropriate and adequate privacy safeguards are inserted into the law in a manner that protects against the inappropriate use of data, or to seek to prevent privacy or security breaches from taking place.

For example, new TNC laws, if not invalidated or repealed for other reasons, should insert new provisions that: (1) impose restrictions on access to data internally and to private third parties without express permission from passengers as to the specific entity or purpose for which such data will be used; (2) create security safeguards imposed and monitored by regulators to ensure that hackers cannot access such TNC data; and (3) stipulate a requirement, as exists in San Francisco and New York City as well as in various Australian states and elsewhere, for the companies doing business with TNCs or TNCs themselves to submit electronic trip sheet data while on-duty, pick-up, and drop-off, as well as fare box data at a minimum. By so doing, regulators can ensure compliance with various laws, and analyze industry economics with a solid factual basis.

---

[100]In the FTC's seminal 2012 report, Protecting Consumer Privacy in an Era of Rapid Change ("2012 Privacy Report"), the Commission made plain its "particular concerns of location data in the mobile context" and called on "entities involved in the mobile ecosystem to work together to establish standards that address data collection, transfer, use, and disposal, particularly for location data." (http://www.ftc.gov/sites/default/files/documents/reports/federal-trade-commission-report-protecting-consumer-privacy-era-rapid-change-recommendations/120326privacyreport. pdf.) Since then, the FTC has issued further guidelines advising mobility app companies on best practices with respect to the development of privacy policies and practices (http://www. business.ftc.gov/documents/bus81-marketing-your-mobile-app; http://www.business.ftc.gov/ documents/bus83-mobile-app-developers-start-security; http://www.ftc.gov/sites/default/files/ documents/reports/mobile-privacy-disclosures-building-trust-through-transparency-federal-trade-commission-staff-report/130201mobileprivacyreport.pdf).

3.3 Part III: Ground Transportation and the Future of Big Data

3.3.1 The Emergence of TNCs as Big Data Companies

As discussed above, data is one of the most valuable assets of TNCs. Uber, for instance, knows where the platform's users live, work, eat, travel, stay, and when they do all these things.[101] This data can be sold to third-party companies, thereby creating a new source of revenue for e-hailing companies.

In February 2015, Uber announced its partnership with Starwood Hotels & Resorts.[102] The reward program allows Uber riders who are Starwood Preferred Guest ("SPG") members to earn "Starpoints" during any Uber ride. To participate, Uber users who are new or existing SPG members need to link their SPG and Uber accounts. After completing a qualifying stay at a Starwood hotel, SPG members can start earning one "Starpoint" for every $1 spent on Uber, and additional bonus "Starpoints" can be accumulated during Starwood hotel stays. In return, customers who opt in let Uber share with Starwood Uber ride data, including the passenger's name, email, photo, pick-up and drop-off locations and times, fare amounts, distances traveled, and Uber products used.[103]

3.3.2 Use of Data by Regulators and Academia

Transportation data has been used by regulators and academia for a variety of purposes, most notably for planning and regulation as well as studies related to real-time operation. An example of the former includes a study conducted by the TLC in 2013 to determine the best location to install electric vehicle charging stations based on demand and supply factors. On the demand side, the level of taxi pick-ups and drop-offs pointed to Manhattan and Western Queens as ideal locations (see Fig. 3.4). On the supply side, it was found that NYC's grid capacity could very well support 350 quick chargers, but with significant constraints in certain parts such as West/Central Midtown and Long Island City.[104]

Using NYC taxi trip data, researchers from Purdue University analyzed spatial and temporal patterns of taxi demand, potential unbalanced trip pattern, intrinsic taxi trip classes (using two-step clustering algorithm) and taxi trip mobility (based

---

[101]Ron Hirson, *Uber: The Big Data Company*, Forbes, March 23, 2015, https://www.forbes.com/sites/ronhirson/2015/03/23/uber-the-big-data-company/#3f3899318c7f.

[102]*See* https://newsroom.uber.com/revving-up-rider-rewards-with-starwood-hotels/.

[103]Ron Hirson, *Uber: The Big Data Company*, Forbes, March 23, 2015, https://www.forbes.com/sites/ronhirson/2015/03/23/uber-the-big-data-company/#3f3899318c7f.

[104]http://www.nyc.gov/html/tlc/downloads/pdf/electric_taxi_task_force_report_20131231.pdf.

Fig. 3.4 Chargers allocation based on taxi activity. Source: http://www.nyc.gov/html/tlc/downloads/pdf/electric_taxi_task_force_report_20131231.pdf. (Note: Color key not available from Source)

on overall travel distance distribution). The study concluded the commonality of unbalanced taxi trips with urban boundaries being a major impediment factor for mobility.[105] This finding was confirmed by another study using Geographically Weighted Regression ("GWR") to model spatial heterogeneity of taxi ridership. In addition, the study also found that medium income level reduces the number

---

[105]Xinwu Qian, Xianyuan Zhan and Satish V. Ukkusuri, *Characterizing Urban Dynamics Using Large Scale Taxicab Data*, http://www.springer.com/cda/content/document/cda_downloaddocument/9783319183190-c2.pdf?SGWID=0-0-45-1508574-p177357756.

of taxi trips at particular places, and established the positive correlation between accessibility to subways and taxi ridership.[106]

Another study examined the labor supply of taxicab drivers using their work hours and collected fares. Using T-PEP data of every taxi trip in January, April, July, and October of 2013, the study provided empirical support of the income-targeting hypothesis and found January as the month when the behavior was most evident. This observation would not have been possible without big data, as limited data could significantly bias the analysis. The TLC's T-PEP data provided a unique opportunity to conduct empirical analyses of labor supply theories which contradicted earlier studies.[107]

On matters related to real-time operations, several numerical experiments conducted using NYC taxi trip data show the effectiveness of time-of-day ("TOD") pricing in the real world. Specifically, the results suggest that by adopting the approximate dynamic programming ("ADP") approach, TOD pricing may increase daily cab revenue by 10%.[108] Another proposed optimization procedure for dynamic ridesharing systems shows a mean delay per user of about 1 min and reduction of match refusals up to 13%.

3.3.3   Future Uses of Data

Data analysis is central to the development of practical policy solutions. Examples include the identification of traffic patterns and road planning; bridge and road maintenance; and communication of real-time traffic information. Data has been successfully used to develop and implement new policies designed to improve transportation services and public safety.

3.3.3.1   Transportation Planning and Improved Taxi Service

The Boro Taxi Program illustrates how data analysis has the potential to be used to address transportation policy and planning issues, and the need for regulators to have access to stakeholders' data. However, although e-hailing companies have become major players on the transportation scene, they still share little information with the public.

---

[106]Xinwu Qian and Satish V. Ukkusuri, *Spatial variation of the urban taxi ridership using GPS data*, https://www.researchgate.net/publication/273792532_Spatial_variation_of_the_urban_taxi_ridership_using_GPS_data.

[107]Ender Faruk Morgul and Kaan Ozbay, Ph.D., *Revisiting Labor Supply of New York City Taxi Drivers: Empirical Evidence from Large-scale Taxi Data*, http://engineering.nyu.edu/citysmart/trbpaper/15-3331.pdf.

[108]Xinwu Qian and Satish V. Ukkusuri, *Time-of-Day Pricing in Taxi Markets*, https://www.researchgate.net/publication/308308542_Time-of-Day_Pricing_in_Taxi_Markets.

Originally, only yellow taxicabs were permitted to pick up passengers in response to a street hail. However, an analysis of trips using GPS by the TLC revealed that 95% of yellow taxi pick-ups occurred in Manhattan below 96th Street and at John F. Kennedy International Airport and LaGuardia Airport.[109] This resulted in significantly lower access to legal and safe taxi rides for people in outer boroughs and upper Manhattan, as they often had to rely on street pick-ups by liveries or unlicensed vehicles. To fill in the gaps, the Five Borough Taxi Plan was started with the Street Hail Livery program that allows Boro Taxis to pick up street hail passengers in the Bronx, Brooklyn, Queens (except the airports), Staten Island, and northern Manhattan (north of West 110th Street and East 96th Street).

The Boro Taxi program addresses 5 major issues in NYC's taxi and for-hire vehicle industries:

- *Mobility*: Neighborhoods outside Manhattan previously lacked access to legal point-to-point transportation without calling ahead.
- *Car Ownership*: Taxis are a form of car sharing and a well-functioning taxi system helps provide alternatives to car ownership.
- *Service Quality*: The quality of street-hail service available outside Manhattan was inconsistent and exposed passengers to fare haggling and other inconveniences.
- *Passenger Safety*: Many passengers have difficulty differentiating legal liveries from illegal cabs.
- *Illegal Activity*: Passengers outside Manhattan who wanted on-demand service by street hailing had no choice but to rely on drivers who were violating the law.
  Similarly, having access to TNCs' data can enable policy-making to be more effective.

In January 2015, UberX announced that it would start sharing anonymized trip data with the City of Boston on a quarterly basis as part of the company's new national data-sharing policy.[110] This information could have potentially been very helpful in analyzing the net effects of surge pricing in the Boston community.[111] The goal of the agreement was to give Mayor Martin J. Walsh's administration unique insight into how people get around the City of Boston, and assist in the development of the City's transportation policy and planning goals.[112] Unfortunately, Uber's failure to provide useful data has made it difficult to conduct any worthwhile analysis.[113] Uber agreed to hand over all trip data on a quarterly basis, but in

---

[109] *See* http://www.nyc.gov/html/tlc/html/passenger/shl_passenger_background.shtml.

[110] Emily Badger, *Uber offers cities an olive branch: your valuable trip data"* The Washington Post, Jan. 13, 2015, https://www.washingtonpost.com/news/wonk/wp/2015/01/13/uber-offers-cities-an-olive-branch-its-valuable-trip-data/?utm_term=.5e54580cdda2.

[111] Adam Vaccaro, *Highly touted Boston-Uber partnership has not lived up to hype so far*, Boston.com, June 16, 2016, https://www.boston.com/news/business/2016/06/16/bostons-uber-partnership-has-not-lived-up-to-promise.

[112] *Id.*

[113] *Id.*

addition to failing to cooperate at times, the data handed over does not show specifically where riders' trips began or ended.[114] Instead, the pick-up and drop-off locations only provide the zip codes, not the actual address.[115] Because Boston's zip code areas are too large, the current data sets do not allow for analysis of how proximity to public transit affects Uber usage, how a new building affects transportation patterns, or how service in particular neighborhoods has been affected by surge pricing.[116]

The lack of sufficient data on TNCs has made it equally difficult to assess their impact on the environment, which stands as one of the many repercussions of the unbridled growth of TNCs. The number of active vehicles on the streets and the growth of vehicles for the sole purpose of providing for-hire transportation, which will inherently require longer than average vehicle miles, have been a concern for policymakers who seek to improve air quality, reduce pollution, and combat global climate change. Recent epidemiological studies have also shown elevated risks of non-allergic respiratory morbidity, cardiovascular morbidity, cancer, allergies, adverse pregnancy and birth outcomes, and diminished male fertility for drivers, commuters, and individuals living near roadways.[117] The lack of sufficient data to correctly measure the impact of the expansion rate of Uber and other TNCs in many cities has exacerbated the problem. These companies do not provide data to substantiate the claims they make about their success in reducing the number of vehicles on the roads, despite the public representations that their core business is developed based on TNC claims of being "everyone's private driver."[118]

Similarly, because TNCs strictly control their data—and much of the data they release to the public portrays them in a positive light—it is difficult to definitively determine the net effects of surge pricing on the wider transportation industry, its consumers and stakeholders.

3.3.3.2 Law Enforcement

Data can be used to enforce traffic infractions and regulations as well as deter fraud. The enforcement of the recently adopted TLC's Driver Fatigue Rules illustrates this point. Medallion and Boro Taxis are equipped with trip-recording equipment, and such records are transmitted to the TLC on a regular basis. FHV bases transmit records of the for-hire vehicles that they dispatch on a regular basis. TLC will review

---

[114] *See id.* Note: Emails show that the city agreed to the zip code limitations as the agreement was drafted in early 2015.

[115] *Id.*

[116] *Id.*

[117] World Health Organization Europe, *Health effect of transport-related air pollution*, 2005, http://www.euro.who.int/__data/assets/pdf_file/0006/74715/E86650.pdf.

[118] Felix Salmon, *The economics of "everyone's private driver"*, Medium, June 1, 2014, https://medium.com/@felixsalmon/the-economics-of-everyones-private-driver-464bfd730b38#.orq4df9gv.

these trip records after submission to calculate the hours in which a driver is picking up passengers in any 24-h or 7-day period. Trips by a driver who accepts dispatches from multiple bases, or who operates both taxis and FHVs, will be combined to determine the total number of hours worked. Bases will only be responsible for trips that they dispatch, not dispatches that their affiliated drivers accept through other bases or street hails accepted by Boro Taxis.[119]

Outside of the driver fatigue context, this new data will also allow the TLC to better enforce complaints by passengers, pedestrians, and motorists, and will support enforcement against illegal solicitations by FHV drivers.

3.3.3.3 Public Safety

One of the boldest initiatives to come out of New York City Mayor Bill de Blasio's administration is Vision Zero, an ambitious plan to eliminate traffic fatalities.[120] A series of high profile crashes involving the deaths of pedestrians in the beginning of 2014 prompted Mayor de Blasio to quickly fulfill his campaign promise by announcing, on January 15, 2014, the creation of the Vision Zero task force comprised of representatives from the New York City Police Department, the New York City Department of Transportation ("NYC DOT"), the TLC and NYC Department of Health and Mental Hygiene. The task force was charged with developing a comprehensive Vision Zero roadmap to eliminate deadly traffic crashes, especially those involving pedestrians. In a little over a month, on February 18, 2014, the task force developed an action plan with 63 recommendations to reduce traffic deaths. The action plan contains proposals for several City agencies, and includes several state and city legislative initiatives. Working in partnership with the Mayor, the New York City Council adopted an historic package of 11 bills, and six resolutions to help implement Vision Zero.[121]

Many of these initiatives are implemented with new technology that helps correct human errors and enforce laws. Vision Zero is not a new idea. Sweden first developed the idea of a Vision Zero Plan in 1995. By 1997, Sweden had adopted legislation to implement the goals of eliminating traffic fatalities and serious deaths

---

[119] See http://rules.cityofnewyork.us/content/driver-fatigue-rules.

[120] See Vision Zero: A Technology, Legal and Policy Overview (New York City & Beyond); TLC Magazine; Black Car News, June 2014; available at: http://www.tlc-mag.com/archive_issues/in_focus_july14.html.

[121] The bills require the City to, among other things: study left hand turns and how to make arterial streets safer as part of a study of pedestrian fatalities and serious injuries due every 5 years; create at least 7 slow zones and 50 slow zones around schools annually, and report on these slow zones annually; suspend the license of any taxi driver who receives a summons for causing a critical injury or death, as well as lifting the suspension of a driver if cleared of charges, and to revoke the driver's license if found guilty; make failure to yield to a pedestrian or bicyclist a traffic infraction, as well as make contact with a pedestrian crossing the street a misdemeanor unless the pedestrian initiated contact; and create enhanced penalties for dangerous taxi and for-hire vehicle drivers.

by 2020, and implemented several initiatives including annual evaluations of road data. Vision Zero would not be possible without analysis of data involving traffic and crashes. With TNCs logging more miles every year, it is important to have their trip data to support any Vision Zero initiatives.

The TLC adopted the Vehicle Safety Technology Pilot as a Vision Zero initiative in 2014. This program involves the use of black boxes to reduce dangerous driving habits such as speeding, distracted, aggressive, or erratic driving that may lead to collisions.[122] Black boxes are different than Event Data Recorders ("EDRs") because they record data continuously. EDRs, on the other hand, only collect data over a brief period of time, specifically during collisions.[123] Along with GPS units, cameras, and accelerometers, black boxes collect telematics such as speed, G-force, braking and acceleration patterns. The data is used to warn drivers who have surpassed speed limits and also to disable taxi meters.[124] Besides the drivers, telematics can also be accessed by fleet managers or vehicle owners through online portals or the cloud. In the aftermath of crashes, black boxes can provide crucial information on the causes of accidents such as hard braking, hard acceleration, hard turning, and abrupt lane changes.[125] This data is far more accurate than traditional diagnostics for accidents based on skid marks and steel deformation.[126] Importantly, the data can be used for investigation of insurance claims, identification of driving patterns or alleged criminal activity, and traffic law enforcement.

3.4 Conclusion and Recommendations: A Need for Third Party Validation

3.4.1 Deficiencies of the TNCs' Self-Regulation Model

The essence of Transportation Network Company ("TNC") laws revolves around a "we can do it faster and better than government" attitude. TNC laws generally transfer the responsibility of conducting background checks and vehicle inspections with less stringent requirements from regulators to the TNC so that they can sign up as many drivers as possible. However, there is an ulterior motive, as no app-based dispatch model works without having an adequate supply of drivers. It is too

---

[122]See http://www.nyc.gov/html/tlc/html/industry/veh_safety_tech_pilot_program.shtml.

[123]See http://www.nyc.gov/html/tlc/downloads/pdf/tlc_black_box_rfi_final.pdf.

[124]Jonathan Lemire, *New York City Council passes Vision Zero legislation*, The Associated Press, May 29, 2014, http://abc7ny.com/traffic/nyc-council-passes-vision-zero-traffic-safety-legislation/83340/.

[125]See http://www.nyc.gov/html/tlc/downloads/pdf/second_vehicle_safety_technology_report.pdf.

[126]Barry Nalebuff and Ian Ayres, *Why Not?: How to Use Everyday Ingenuity to Solve Problems Big and Small*, Harvard Business School Press, 2003.

costly and difficult to entice and subsidize the transfer of professionally licensed black car and taxicab drivers to TNCs (although this was done at a very high cost by Uber in New York City). TNCs claim that college students and part-time workers would be discouraged by the process of purchasing insurance, completing physical paperwork, leaving their homes or computers and undertaking a simple 5 min fingerprint check. While there is some truth to the convenience factor, the motive of TNCs is to attract more drivers by making it easier for drivers to become approved as TNC drivers while managing and assuming the risks of some potentially unsafe or inexperienced drivers who slip through the cracks and cause harm to others. The self-regulation model allows the TNCs to control the information pertaining to public incidents such as sexual assaults and crashes, and discourages further media coverage as such information—which is of public interest—has been labeled as "proprietary" and not subject to public disclosure laws that keep such stories and criticism alive. The self-regulation model is an effort to control information, make licensing more efficient and to facilitate a less costly market takeover.

It is possible to engage in modified self-regulation of transportation companies, as is done with trucking and limousine companies engaged in interstate commerce by the U.S. Department of Transportation's Federal Motor Carrier Safety Administration ("FMCSA"). The FMCSA requires that interstate truckers and drivers obtain medical exams[127] and not work more than a certain number of hours during a time period for public safety reasons.[128] FMCSA-licensed carriers must collect information ensured by Federal auditor compliance.[129] As such, the key ingredients for the success of a self-regulation system are auditing resources and significant penalties and fines to serve as an appropriate and effective deterrent. Without unbridled access to TNC data to audit real-time performance and compliance at all levels, including the ability to impose significant fines, this model is doomed to fail. In general, governments, and not private parties, should be the ones who regulate; but if TNCs are allowed to engage in self-regulation due to the lack of government resources, they should be required to pay for enforcement resources and provide their data for auditing and compliance.

3.4.2 Enforcement of TNC Laws

Since the advent of disruptive for-hire ground transportation app-based TNCs, such as Uber and Lyft, many laws have been enacted throughout the United States to permit market entry of TNCs through partial deregulation and self-regulation of these services, mostly at a State level after facing opposition in many localities

---

[127]See https://www.fmcsa.dot.gov/faq/Medical-Requirements.

[128]See https://www.fmcsa.dot.gov/regulations/hours-of-service.

[129]See   https://www.fmcsa.dot.gov/international-programs/certification-safety-auditors-safety-investigators-and-drivervehicle.

("TNC laws"). TNCs have, for the most part, pushed for statewide regulation and laws that semi-legitimize their business model, because doing so at the state level involves conceivably less lobbying, legal, and media-related resources than engagement in a much larger number of municipalities, counties, and/or villages. TNCs have recognized that states typically lack the same enforcement capabilities as localities, and it is no mistake that state regulation has been sought.

As of March 2017, 43 States and the District of Columbia have passed some sort of TNC legislation.[130] These TNC laws define a TNC by describing the same exact activity or service performed by taxicabs and limousines—transporting a paying passenger from point A to point B—and specifically exempt incumbent operators.[131] Several of the newly enacted laws are "cookie cutter" versions of one another, which provide modest standards with which TNCs must comply in the areas of: licensing; insurance; driver vetting; vehicle standards; and accessibility.

For example, in California, some TNC regulatory requirements include:

• Conducting, or have a third party conduct, a local and national criminal background check for each participating driver that shall include both a multistate and multi-jurisdiction criminal records locator or other similar commercial nationwide database with validation; and a search of the United States Department of Justice National Sex Offender Public Web site.[132]
• Inspecting all vehicles and maintaining the records of such inspections in case of an audit.[133]

In Colorado, TNC regulatory requirements include:[134]

• Obtaining and reviewing a criminal history record check for the individual before permitting an individual to act as a driver on their digital network;
• Maintaining copies of TNC medical examiner's certificates for all TNC drivers that are authorized to access its digital platform;

---

[130] See https://tti.tamu.edu/policy/technology/tnc-legislation/.

[131] For example, although many taxicab and limousine companies utilize a digital platform such as a website or App allowing passengers to book trips, Tennessee defines a TNC as "a business entity operating in this state that uses a digital network to connect riders to TNC services by TNC drivers" and explicitly distinguishes a TNC "from a taxi service, limousine service, shuttle service, or any other private passenger transportation services that are regulated pursuant to present law." See https://trackbill.com/bill/tn-hb992-transportation-dept-of-as-enacted-enacts-the-transportation-network-company-services-act-amends-tca-title-7-title-54-title-55-title-56-and-title-65/1135990/.

[132] See Assembly Bill No. 1289, Chapter 740, approved by the Governor on September 28, 2016, available at: https://leginfo.legislature.ca.gov/faces/billTextClient.xhtml?bill_id=201520160AB1289.

[133] State of California Public Utilities Commission, *Transportation License Section State of California Public Utilities Commission Basic Information for Transportation Network Companies and Applicants*, p. 4, http://www.cpuc.ca.gov/uploadedfiles/cpuc_public_website/content/licensing/transportation_network_companies/basicinformationfortncs_7615.pdf.

[134] See Code of Colorado Regulations, Sections 6700, *et seq.* https://drive.google.com/file/d/0B8qvU2knU8BkRHhad0EwZVVuVTA/view.

- Maintaining the following data for each prearranged ride, as applicable, for a minimum of 1 year from the date of each such prearranged ride: the personal vehicle's license plate number; the identity of the driver; the identity of the matched individual using the TNC application to request a prearranged ride; the date and time of the rider's request for service; the originating address; the date and time of pick-up; the destination address; and the date and time of drop-off;
- Conducting or having a vehicle inspector conduct an initial safety inspection of a prospective driver's vehicle before it is approved for use as a TNC vehicle and at least annual periodic inspections of TNC vehicles;
- Adopting a policy designed to ensure that, after 16 cumulative hours logged into the TNC's digital network in a calendar day, the driver shall log out of the TNC's digital network for eight consecutive hours; enforcing this policy through appropriate monitoring of available data and administration of disciplinary actions;
- Adopting a policy designed to ensure that no driver is logged in to the TNC's digital network for more than 70 h in a consecutive 7-day period; enforcing this policy through appropriate monitoring of available data and administration of disciplinary actions.

This self-regulatory TNC framework has been exposed as having many deficiencies. In April 2017, Massachusetts officials disqualified over 8000 Uber, Lyft, and other app-based vehicle drivers for failing a criminal background check.[135] According to the report: "Hundreds were disqualified for having serious crimes on their record, including violent or sexual offenses, and others for driving-related offenses, such as drunken driving or reckless driving, according to the state Department of Public Utilities".[136] Also disqualified were 51 sex offenders and 352 offenders in incidents related to "Sex, Abuse, and Exploitation."[137] Despite the inferior background checks TNCs conduct, a plethora of convicts of criminal offenses were also recently discovered in Boston and Houston.[138]

3.4.3   The Solution of a Third-Party Validator

A third-party validator hired by the government and paid for by the TNCs to collect, maintain, and analyze data is the solution to the different issues raised in this chapter.

---

[135] Adam Vaccaro and Dan Adams, *Thousands of current Uber, Lyft drivers fail new background checks*, The Boston Globe, April 5, 2017, http://www.bostonglobe.com/business/2017/04/05/uber-lyft-ride-hailing-drivers-fail-new-background-checks/aX3pQy6Q0pJvbtKZKw9fON/story.html.
[136] *Id.*
[137] *Id.*
[138] *Id.*; Joel Eisenbaum, *Houston mayor: 50 percent of Uber driver applicants have criminal record*, Click2Houston, March 30, 2017, http://www.click2houston.com/news/investigates/houston-mayor-50-percent-of-uber-driver-applicants-have-criminal-record.

A third-party validator would collect, monitor, and audit the aforementioned items including, but not limited to, granular pick-up and drop-off locations and times, collision or "black box" data, and duration of trips. A third-party validator would be able to test data accuracy, protect trade secrets, provide transparency, and assist with law enforcement, and enable regulators to access this information under conditions acceptable to the TNCs to assess market conditions and make policy decisions.

In general, governments, and not private parties, should be in charge of regulating the ground transportation industry. However, since TNCs are allowed to self-regulate in many jurisdictions and due to the lack of government resources, they should be required to pay for a third-party validator to ensure their data is accurate and turn over their data to facilitate auditing and compliance. As such, the key ingredients for the success of a self-regulation system are auditing resources, and significant penalties and fines to serve as an appropriate and effective deterrent. Without unbridled access to TNC data to audit real-time performance and compliance at all levels, including the ability to impose significant fines, public safety is jeopardized and regulators' responsibilities to make policy decisions hampered.

Due to the lack of either appropriate TNC standards or any commitment to enforcement at the local level, TNCs should be required to provide data in an anonymized format or lockbox via an approved third party administrator hired by the government. The law can create an exemption from Freedom of Information Laws ("FOIL")[139]—which could otherwise be an open platform for public use of the data—and allow access exclusively to government regulators for specific investigatory or data collection purposes that are clearly defined (i.e., for fare increases; traffic or environmental studies; to investigate crimes and complaints; or to return lost property). There is a precedent where some states have made FOIL exemptions for information received from financial firms that manage the public pension funds, finding that such information is proprietary.

It is imperative that third-party validators are competent in big data analytics, specifically data mining, machine learning, and predictive modeling. Importantly, these validators should have successful track records in providing third-party auditing and verification of state transportation agencies and/or private companies that are involved in data analytics and data management. In the ideal scenario, they should also have sufficient experience in managing information systems related to the Internet of Things ("IoT") and autonomous vehicles ("AV"). The IoT and AV are some of the emerging technologies within the intelligent transportation systems ("ITS"), which would bring about an increasing volume, velocity and veracity of ground transportation data.

---

[139]FOIL laws usually have provisions for information determined to be exempt from disclosure for public policy reasons. For instance, New York State's FOIL Law contains exemptions for certain information including information that if disclosed would constitute an unwarranted invasion of personal privacy. (*See* NYS Public Officers Law §87(2)). The newly adopted New York State TNC law exempts from public disclosure the names and identifying information of TNC drivers obtained for an audit (*See* New York State Vehicle and Traffic Law §1698 2., effective June 29, 2017).

We are on a collision course between private monetization and control of ground transportation data, and the desire for government regulators and agencies to access such data for compliance and planning purposes. There is a trend in cities to create open data platforms to not only study smart city data point in the academic research and government policy arena, but also to provide such data to private individuals and companies to spur innovation, including the creation of smartphone applications and new technological products and services. Universities and academic researchers also crave ground transportation data for the purpose of study and analysis to publish papers that offer new solutions to longstanding transit problems.

The desire for private companies, as part of their business model, to protect intellectual property rights and maintain valuations to withhold this information from their competitors, the government and academics, is a topic for debate which will continue for some time. Due to Federal and State Sunshine Laws (or Open Government Laws such as the Freedom of Information Laws enacted following the Nixon-Watergate scandal of the 1970s), providing such data to the government may ensure that any member of the public can access this now public data.

The solution to this conflict could very well be laws that exempt such data from sunshine laws, to preserve personal privacy. Or, more appropriately, to allow data to be anonymized, collected, and aggregated by a third party private vendor, with contracts to ensure data security and privacy, and to allow a lockbox or data commons to be accessed by the public, the government and researchers. This would sidestep concerns about the disclosure of trade secrets to competitors and would be a mechanism to outsource auditing functions, oversight and regulation of private ground transportation carriers.

Chapter 4
A Privacy-Preserving Urban Traffic Estimation System

Tian Lei, Alexander Minbaev, and Christian G. Claudel

4.1 Introduction

Traffic congestion is an increasing concern in large urban areas of the world, and is expected to become worse as global traffic demand increases. With ever increasing societal costs, traffic congestion can be addressed through a variety of methods, including planning, construction of additional capacity on the transportation network, and traffic control methods. Among all these methods, traffic control have one of the greatest cost/benefit ratio since they do not require major construction or relocation, and, unlike planning, have an immediate impact. Several traffic control techniques have been successfully tested in the past, including ramp metering [29, 31], adaptive speed limits [21, 26], demand response [3], or boundary control [18, 23, 24]. However, all these methods require as an input accurate traffic density, velocity and flow estimates. Uncertain traffic estimates can lead to poor performance of the closed loop control system, and if the uncertainty is high enough, the performance of the closed loop system may be worse than when using no control at all.

Hence, accurately monitoring traffic in the region of interest is of critical importance. Monitoring traffic requires the fusion of measurement data with sensors, which can be either fixed or mobile (moving alongside traffic). The first category is usually expensive to deploy and maintain, and is increasingly replaced by mobile

T. Lei
University of Texas, Austin, TX, USA

Chang'an University, Xi'an, China

A. Minbaev · C. G. Claudel (✉)
University of Texas, Austin, TX, USA
e-mail: alexander.minbaev@utexas.edu; christian.claudel@utexas.edu

© Springer International Publishing AG, part of Springer Nature 2019
S. V. Ukkusuri, C. Yang (eds.), *Transportation Analytics in the Era of Big Data*,
Complex Networks and Dynamic Systems 4,
https://doi.org/10.1007/978-3-319-75862-6_4

traffic sensors, also known as probe vehicles. In the recent years, these vehicles, containing speed and position sensors, have emerged as a possible solution to the problem of monitoring traffic flow. Probe sensing has a very low marginal cost, particularly when sensing relies on existing devices (for instance smartphones), see, for instance, http://traffic.berkeley.edu/. Nevertheless, all current probe-based traffic monitoring systems require users to send their location data to a centralized server, which carries high risks of user privacy intrusion whenever the location data servers are attacked. Current probe-based traffic sensing systems also rely on satellite-based positioning (or cellular-based positioning), which is privacy intrusive by nature, since it reveals the approximate position of users. It should be noted that even anonymous location tracks can yield substantial information on users [20], which can be correlated with social network data to identify user identity based on their tracks. While *Basic Safety Messages* (BSMs) sent over *Dedicated Short Range Communication* (DSRC) will probably be mandated soon [37], these messages will only be available to users around a specific vehicle, and carry less risks of privacy intrusion.

In this chapter, our objective is to investigate a new type of traffic monitoring system built around existing WiFi or Bluetooth readers, and capable of maintaining the privacy of users, while providing accurate traffic data with low penetration rates. All techniques that are currently used to enforce user privacy are based on either a modification of the sampling characteristics [15] (locations of samples, sampling rate) of the probe sensor or rely on obfuscation of the actual location track of the user by removing data points or adding fake data points to it. A spatial sampling method called *virtual trip lines* (VTLs) is proposed in [17], to prevent users from sending their data whenever they are close to locations that could help identify them (home, workplace). However, this method is not applicable for traffic monitoring in urban environments since most urban areas are either workplaces or accommodations. Another obfuscation method is shown in [30], but the same chapter shows that generating fake data to hide real location tracks is challenging, even with aggregated statistical data. Ultimately, the obfuscation techniques only increase the level of privacy of a user, but do not guarantee it. In the worst case scenario of a single user sending data over a given region, it is pretty easy for an attacker to identify the user path. Privacy enforcing algorithms tend to eliminate altogether these location tracks, though, from a traffic monitoring point of view, these isolated tracks are also the most valuable source of data, since they provide data in locations where no other measurement data is available.

In [8], we investigated a system based on short range communication of probe data to a wireless sensing infrastructure. In this chapter, we propose a similar approach, though we rely on IMUs for the probe vehicles, and therefore do not generate absolute position measurements. The wireless sensing infrastructure does not only estimate the state of traffic to create traffic maps (as in [8]), but is also used to determine the trajectory of a user, based on trajectory estimates obtained by integration of the IMU measurements.

The rest of this chapter is organized as follows. We first present an overview of the complete system in Sect. 4.2. We then describe the IMU component of the system in Sect. 4.4, and describe specific issues associated with the calibration and trajectory estimation in Sect. 4.5. Finally, we present in Sect. 4.6 a possible implementation of this system, including a distributed computing approach for traffic state estimation over a transportation network.

4.2   System Overview

4.2.1   Current Architecture of Probe-Based Traffic Sensing Systems

Probe-based traffic sensing systems follow typical sensor network architectures, in which data generated by sensors is sent to a centralized server for processing or display [36]. Traffic speed and/or density maps are the end product for the user, and the basis of all other location-based services such as travel time estimation or optimal routing. One of the major drawbacks of such systems is the fact that the ID proxy server holds privacy sensitive information regarding the users. The ID proxy server is a vulnerable target, since it handles personally identifiable information (for example, cellphone number), and removes it from the data. Nevertheless, the privacy of users is at risk even when data is anonymized [19], and input databases (which are used to estimate the state of traffic) can also be a vulnerable target, as vehicles can be tracked using traffic flow models, as in [6, 7]. The architecture of typical probe-based traffic monitoring systems is illustrated in Fig. 4.1.

Some systems [36] attempt to solve the privacy problem using data obfuscation or specific spatial sampling strategies [17]. However, none of these strategies can guarantee that user privacy is preserved in all situations. In a worst-case situation, no sampling strategy can prevent someone to re-identify the approximate path of

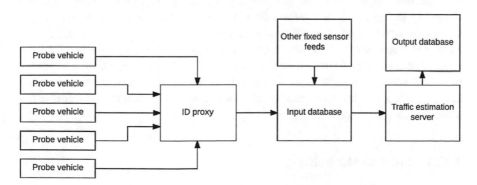

Fig. 4.1 Architecture of current probe-based traffic monitoring system. The data generated by probe vehicles is anonymized by an ID proxy, and the corresponding data is sent to an input database, which also receives data from other (fixed) sensors. The data is the processed by a central traffic estimation server, and sent to an output database for dissemination

a single user evolving in a given geographical area, unless no data is transmitted at all. Another issue is that obfuscation and sampling strategies are degrading the performance of the system, since they lower the accuracy of the input data. This issue is further compounded by the fact that not all data is equal. The extent to which a sample of traffic data reduces the uncertainty of the state of the transportation network depends on the average density of data points around this particular sample. Paradoxically, the privacy strategies outlined above degrade the quantity and quality of input data in areas in which this data would be most valuable, that is areas in which almost no data exists. Therefore, imposing privacy constraints on a traffic monitoring system can have a very strong impact on the quality and reliability of traffic estimates.

Since user location and velocity information are required by the model to build the traffic maps through estimation, the problem of user privacy can only be solved through decentralized estimation, in which data is only exchanged locally. Indeed, a centralized server handing user data can always be a target of attacks. A centralized traffic monitoring system also requires long range communications, which also requires some authentication, and thus contains privacy sensitive data. Hence, the system should not require long range communication, and should distribute the processing among some computational nodes.

4.2.2 Proposed System Architecture

Since Bluetooth and WiFi readers are commonly used in traffic flow monitoring, the proposed system relies on these devices as endpoints and traffic computing tools. The readers are scanning all corresponding devices (Bluetooth or WiFi) located in their radius of detection. The devices have to transmit traffic data to the readers during this scanning process. Scans do not allow direct connections between devices, and therefore the only mechanism of data transmission from the device to the reader is the information transmitted by the device to the reader during the scan process. In the Bluetooth protocol, this information consists in a MAC (Medium Access Control) address, and possibly of a device name. The total amount of information exchanged corresponds to 254 bytes, including 6 bytes for the MAC address and 248 bytes for the device name. Since the device name field can transmit a much larger quantity of information, we will transmit information through a modification of this field.

4.2.2.1 Fixed Reader Nodes

The reader nodes play three roles: communication of the traffic state estimate maps to a centralized server, computation (vehicle trajectory estimation, traffic sate estimation) and sensing (through classical Bluetooth or WiFi MAC address re-identification, if needed.[1]) The reader nodes do not have to form a mesh network: traffic readers are usually connected to a central server through a cellular or wire connection. In future implementations of this system, DSRC readers can be used in lieu of `Bluetooth` or `WiFi` readers to collect traffic data and perform the required computations.

4.2.2.2   Probe Sensors

The system we propose is based on measurements generated by an Inertial Measurement Unit (IMU) located inside a vehicle. IMUs consist in a combination of accelerometers and gyroscopes, with possibly a magnetometer for heading estimation. The accelerometers and gyroscopes monitor the accelerations and rotation rates experienced by the vehicle. These measurements can be used to estimate the trajectory of the probe vehicle, and to derive features associated with traffic, for example using machine learning to recognize specific patterns. These patterns include stop and go waves in congestion, very slow continuous traffic, regular stops at traffic lights.

4.2.2.3   Principle of Operation

The network of fixed reader nodes is partitioned into clusters, for example based on Voronoi diagrams. Each cluster consists in a single `Bluetooth` or `WiFi` reader. All clusters need to exchange information with nearby clusters in the form of boundary condition estimates. These information exchange does not contain any user data, and is thus not affecting the privacy of users.

Probe vehicles equipped with IMUs continuously estimate their trajectories, as in Sect. 4.5. These trajectory estimates are encoded as the device name (using a given encoding format). Additional traffic features are inferred along the path using a supervised learning approach. These features can include the presence of stop and go waves, the classification of the type of stops (intersection, traffic, traffic light or curb stop), and the presence of disruptions on the road network, which can be sensed through gyroscope measurements when the driver steers their car in opposite directions to avoid debris or a disabled vehicle.

The trajectory estimates and features are then used to estimate the state of traffic, in a decentralized way. The traffic state estimation process relies on traffic measurement data generated alongside the vehicle trajectories, possibly augmented by fixed sensor data (for example, generated by traffic cameras, radars, loop detectors) if available. All these measurements are integrated in a network traffic state estimation problem corresponding to each cluster. Solving each independent

---

[1]The use of the readers in their usual configuration (re-identification of MAC addresses) requires the transmission of MAC addresses within nodes, and has potential for privacy intrusion.

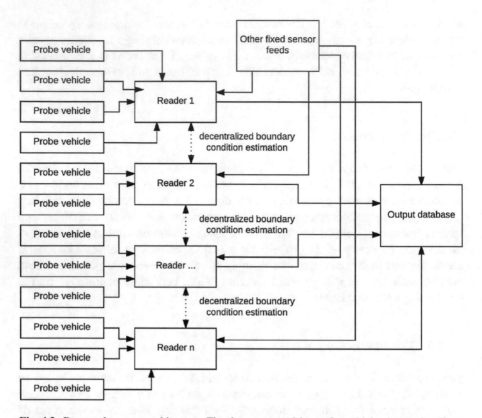

Fig. 4.2 Proposed system architecture. The data generated by probe vehicles is sent locally to each reader, which handles a given sub-network. The data generated by these probe vehicles is integrated with (possibly) fixed sensor data to generate traffic estimates. These estimates are then regularized between sub-networks through consensus-type filtering, and sent to an output database by each reader

state estimation problem yields estimated flows at the boundaries of each cluster. An iterative approach (for example, based on consensus filtering) can be used to make these estimates match between adjacent clusters. This results in a global traffic map, which is obtained by combining the local maps generated by each cluster. The traffic estimation process is summarized in Fig. 4.2.

While no user information is directly shown in this global traffic map, it may nonetheless reveal user presence in some circumstances, for example if a single user affects the traffic state estimates in a way that would reveal them. Several methods such as differential privacy analysis can be used to analyze the global maps and evaluate the potential for privacy breaches. Note that all map-generating systems would also face the same problem.

4.3   User Privacy Analysis

4.3.1   Threat Model

In this chapter, we assume that attackers could compromise any part of the system, that is, any reader handling a network cluster.

4.3.2   Properties of the System

By construction, no vehicle track information can be obtained beyond the radio range of the vehicle transceiver. Since the technology chosen for this work is based on Bluetooth, WiFi or DSRC, the radio range of the system is quite limited. Thus, an eavesdropper can track the position of a vehicle only if they can listen to all clusters in the path of the vehicle. While such a distributed attack is theoretically possible, it is very costly and impractical, requiring to either compromise the radio scanners, deploy new radio scanner infrastructure, or eavesdrop the radio traffic around the radio scanner in all clusters.

4.3.2.1   Consequences of a Reader Attack

Compromising the software of a node can usually be detected by monitoring the input and output data flows, and is thus not particularly stealthy. Similarly, deploying a radio reader infrastructure in a city would locally create Bluetooth discovery signals at regular intervals, which could be detected. An eavesdropping infrastructure is more discrete, but would still require wired or wireless communication. Both type of communication could be detected, either physically (wires) or through radio scanners (wireless).

The cost of such an attack is similar to deploying a full radio monitoring infrastructure, and therefore such attacks are unlikely to be carried out. An attack over a smaller subset of the network would give partial information over a limited geographical area, but would remain expensive to carry out.

Attacking a single reader would reveal the trajectory information of users around this area, provided that the users are in range of the reader. These trajectories can contain privacy sensitive information. By construction of the system, the average spatial separation between consecutive readers defines a spatial scale at which trajectory information is valuable. Thus, the vehicle IMUs do not need to encode the trajectory information on a distance greater than this scale. This naturally defines a cutoff distance in the trajectory integration problem. This cutoff distance can be chosen to be smaller to minimize the risk of privacy intrusion, at the expense of requiring a larger density of readers.

One of the greatest benefits of this system is that no user information is exchanged between readers. Attacking a reader will allow one to obtain boundary condition estimates from surrounding readers (which are used to estimate the traffic maps), though these boundary condition estimates represent aggregated flow estimates, and do not reveal privacy sensitive information.

4.3.2.2 Consequences of a Database Attack

Since the traffic estimation process is distributed over clusters, there is no need for an input database. The output database contains all information generated by each cluster, in the form of traffic maps. This information is publicly available, and thus the attacker would not gain any additional information from attacking this database.

4.3.2.3 Trajectory Inference from Traffic Maps

Since the traffic maps themselves are estimated from traffic measurement data, there exists a relationship between the inputs $x(\tau)_{\tau \in [0,t]}$ to the traffic estimation system (consisting in sensor measurements, probe data, and estimated demand-supply patterns) and the output $y(t)$ (corresponding to the traffic map at time t). However, inverting this relationship is difficult in practice, even if an attacker perfectly knows the inner mechanisms of the traffic state estimation, for the following reasons:

- The relationship between the inputs and outputs is highly nonlinear in general, resulting from an estimation framework that involves a large number of variables. For example, estimating the input data associated with a *Mixed Integer Linear Programming* (MILP) framework such as the framework investigated in this chapter is challenging, since these elements appear as constraints to the problem. Similarly, estimating the input data associated with an Ensemble Kalman Filter (EnKF) traffic state estimation framework such as in [34, 35] is challenging, since the attacker does not know the initial conditions associated with the ensemble choice (which is updated at each step), nor the random noise that is added to the outputs to perform the update step. These random variables affect the estimated output in an unpredictable way.
- Some processes that run in each cluster can further add to the complexity of inverting the relationship between inputs and outputs. For example, the boundary condition estimates (used to solve the first step of the decentralized optimization problem) can be learned over time, based on measurement data obtained during some past time window. The raw measurement data used as part of these boundary condition estimates cannot be perfectly reconstructed, and therefore the boundary condition estimates are not available to an attacker.
- Some degree of noise can be added to the traffic maps, as in [12]. Note that even without noise, some operations such as the quantification of traffic estimates into discrete traffic states (represented by different colors) causes an inherent loss of information. The temporal discretization of the map updates is similarly causing a loss of information.

4.4   Inertial Measurement Unit Based Traffic Flow Monitoring

We use a custom-developed GPS/IMU system (Fig. 4.3) based on an `Arm Cortex M4` processor operating at 168 MHz. It contains a 9-DOF IMU (accelerometer, gyrometer, and magnetometer) and a GPS,[2] and is powered through a USB port that is rigidly attached to the vehicle (through a car charger or a vehicle USB port), albeit at a random orientation with respect to the coordinates of the vehicle. The device can send data over a `IEEE 802.15.4 XBee` module transceiver or a `Bluetooth` transceiver (which is used in this study), at a 10 Hz rate.

A smaller version of this system is shown in Fig. 4.4 below. This version omits the SD card reader and the GPS module, and is more compact (in addition to not generating traffic measurement data).

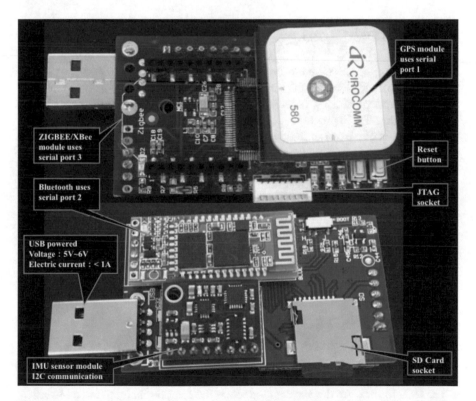

Fig. 4.3 Custom-developed IMU board with Bluetooth module. This module consists in a combination of an IMU (with accelerometer, gyroscope, and magnetometer), a Bluetooth transceiver, a USB port (for power only), an SD card reader, and a GPS module for validation. All peripherals are connected to an `Arm Cortex M4` microcontroller

---

[2]In the present chapter, the GPS data is only used for validation.

Fig. 4.4 Second generation IMU board. This module (right) consists in a combination of an IMU, a Bluetooth transceiver, a USB port (for power only). The earlier prototype version is shown on the left for comparison

4.5 IMU Calibration and Trajectory Estimation

The proposed system is based in IMUs equipped vehicles (with GPS data used only for validation). Generating traffic measurement data from IMUs is nontrivial, and requires several processing steps at the vehicular sensor level and in the wireless sensor network. These processes are highlighted in Fig. 4.5. The first step is to map the coordinates of the sensor to the coordinates of the vehicle, which will be referred to as automatic calibration in the remainder of the chapter. The resulting acceleration and rotation rate measurements from the sensor are mapped into the coordinates of the vehicle, and are used to determine the orientation of the vehicle with respect to the Earth. This allows us to compute the coordinate acceleration (in the Earth frame) by canceling the gravitational component of the acceleration. We then use the acceleration and rotation rate measurements to both estimate the yaw angle and the actual vehicle velocity. This combination allows us to generate vehicle trajectories, which are then encoded and transmitted to the readers during the discovery process.

4.5.1 Automatic IMU Calibration

Unlike GPSs, the orientation of an IMU sensor has an importance. To determine the trajectory of the vehicle in the Earth frame, it is critical to determine the orientation of the IMU to measure the acceleration along the longitudinal, lateral, and vertical axes of the vehicle. This can be achieved by carefully determining the orientation of the device in the vehicle (assumed to be constant, since the device is rigidly connected to an USB port), and compute a corresponding rotation matrix mapping the coordinates of the device to the coordinates of the vehicle.

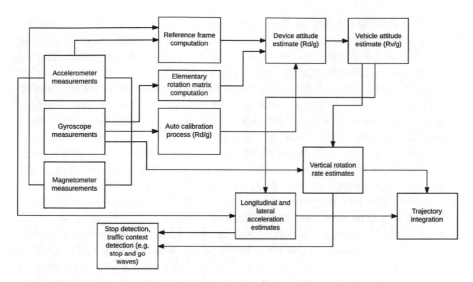

Fig. 4.5 Trajectory estimation and traffic measurement overview. This figure summarizes the steps required to estimate vehicle trajectories and generate traffic measurement data using IMU measurements

To facilitate the deployments, we developed an automatic calibration procedure, detailed in [28]. This procedure allows the IMU to identify its orientation with respect to the vehicle automatically, after a few minutes of driving.

We now focus on reconstructing the trajectory of the vehicle in the Earth frame. This process requires the computation of the attitude of the vehicle (that is, the orientation of the vehicle with respect to the Earth frame), since the trajectory is derived from the integration of the coordinate acceleration, whereas an accelerometer measures the proper acceleration, that is, the acceleration of the vehicle with respect to a free-falling frame. The coordinate acceleration can be computed from the proper acceleration using the formula $a_c = a_p - g$, where g is the vector of acceleration due to gravity.

4.5.2   Attitude Estimation Algorithms

The attitude of the vehicle is encoded by a rotation matrix that translates the vehicles coordinates to the Earth coordinates. This rotation matrix can be estimated using a Kalman Filter [22], or a complementary filter such as the direction cosine matrix (DCM) outlined below. The latter is generally used in attitude estimation and control of ground or air vehicles, as in [32]. This filter is based on the following assumptions:

- The gyroscopes are used as the primary source of orientation information. For short time horizons, we assume that the rotation of the object is small, and the rotation of the frame between times t and $t + \Delta t$ can be represented by the elementary rotation matrix $R_{g,t,t+\Delta t}$, outlined below.
- The DCM filter relies on a rotation matrix estimate generated by the accelerometer and the magnetometer, and integrates the measurements of the gyroscope and the current estimate of the attitude into a new estimate. Since the accelerometer and magnetometer reference matrix represent the low frequency component of the signal, and the rotation matrix $R_{g,t,t+\Delta t}$ represents the high frequency component of the signal, we use the update equation $R(t + \Delta t) = \lambda R_{g,t,t+\Delta t} \times R(t) + (1 - \lambda) R_{\text{ref}}(t)$, where $R_{\text{ref}}(t)$ is the rotation matrix obtained from the accelerometer and magnetometer measurements. The coefficient λ is related to the time constant of the complementary filter, and can be used to place a larger weight on the gyroscope or accelerometer/magnetometer measurements. Its value is a function of the noise levels of the accelerometer and gyroscope, and is very close to one in practice. A large value of λ puts more weight in the gyroscope measurements, which can lead to static positioning errors due to gyro drift. A small value of λ gives too much weight to the accelerometer and magnetometer measurements, which are not perfect attitude reference vectors (due to magnetic perturbations or vehicle longitudinal and lateral accelerations).

The elementary rotation matrix $R_{g,t,t+\Delta t}$ is given by:

$$R_{g,t,t+\Delta t} = \begin{bmatrix} 1 & -g_z \Delta t & g_y \Delta t \\ g_z \Delta t & 1 & -g_x \Delta t \\ -g_y \Delta t & g_x \Delta t & 1 \end{bmatrix} \tag{4.1}$$

The matrix $R_{\text{ref}}(t)$ can, for instance, be determined using the normalized acceleration vector $\frac{a}{\|a\|_2}$ as a first reference vector, and the normalized projection of the magnetic field vector $\mathbf{b}$ on a plane perpendicular to a: $\frac{\mathbf{b} - <\mathbf{b} \cdot \frac{a}{\|a\|_2}> \frac{a}{\|a\|_2}}{\|\mathbf{b} - <\mathbf{b} \cdot \frac{a}{\|a\|_2}> \frac{a}{\|a\|_2}\|_2}$ as a second reference vector. With these reference vectors, the attitude of the vehicle is determined with respect to the up/magnetic North coordinates.

With these results, the coordinate acceleration in the device frame becomes $\mathbf{a_p}$ $(t) - R_{s/g}(t) \begin{bmatrix} 0 \\ 0 \\ g \end{bmatrix}$, and can be used to estimate the trajectory of the vehicle by integrating the acceleration measurements.

4.5.3 Vehicle Trajectory Estimation

The calibration and attitude estimation steps allow us to determine the attitude and the coordinate acceleration of the vehicle (in its frame) at all times. From these, estimating the vehicle trajectory can be done through the integration of the acceleration and attitude measurements. Indeed, for a simple two-dimensional evolution, in which the vehicle is assumed to remain parallel to the mean surface of the Earth, the coordinates of the vehicle $(x(t), y(t))$ in the Earth frame satisfy the following equations:

$$\begin{cases} x(t) = x(t_f) + \int_t^{t_f} v(\tau)cos(\psi(\tau))d\tau \\ y(t) = y(t_f) + \int_t^{t_f} v(\tau)cos(\psi(\tau))d\tau \\ v(t) = v_{(t_f)} + \int_t^{t_f} a_l(\tau)d\tau \\ \psi(t) = \psi(t_f) + \int_t^{t_f} g_z(\tau)d\tau \end{cases} \tag{4.2}$$

where $g_z(\cdot)$ and $a_l(\cdot)$, respectively, represent the yaw rate of the vehicle (vertical component of the gyrometer measurements) and the longitudinal acceleration of the vehicle (component of the acceleration along the longitudinal axis of the vehicle). The coordinates $(x(t_f), y(t_f))$ are approximated as the reader location, given the very short range (tens of meters) of `Bluetooth` or `WiFi` signals.

While the integration of Eq. (4.2) yields the estimated trajectory $(x(t), y(t))$, this numerical process is unstable, and diverges after a relatively short amount of time. To reduce the level of uncertainty associated with the trajectory integration, we use an algorithm to detect stop events, and reset the velocity to zero whenever a stop event is detected. Stop events also allow us to estimate the accelerometer gyrometer bias, which is slowly time-varying, in function of the ambient temperature. These estimates allow a more precise integration to be carried out. An experimental trajectory estimate is shown in Fig. 4.6.

In practice, the trajectory estimates need to be mapped onto the road network. One of the simplest ways to achieve this is to approximate the estimated trajectory into piecewise linear segments, and use the final location of the vehicle (corresponding to the coordinates of the reader) in conjunction with the road network topology to enumerate all paths that could be possibly taken by the vehicle. The most likely path can then be determined by optimizing a cost function (for example a function of the discrepancy between observed distances and turn angles) among all possible paths. This approach has been successfully tested in [27].

This measurement data resulting from the trajectory estimates can then be used as input data for the network traffic estimation framework derived in Sect. 4.6.

4.6 Distributed Computing for Traffic State Estimation

4.6.1 Network Traffic State Estimation

The data assimilation scheme we propose in this article is based on the seminal *Lighthill Whitham Richards* [25] (LWR) traffic flow model, a first order scalar

Fig. 4.6 Illustration of an experimental trajectory integrated using (4.2). The actual trajectory is shown in the upper subfigure, while the estimated trajectory is shown in the lower subfigure. As can be seen from this figure, the integrated distances and angles do not perfectly match the actual vehicle trajectory, though it yields a good idea of the possible vehicle path. This estimated trajectory can be mapped onto the actual road network by leveraging the final position of the vehicle, which can known to be within the range of the `Bluetooth/WiFi` reader

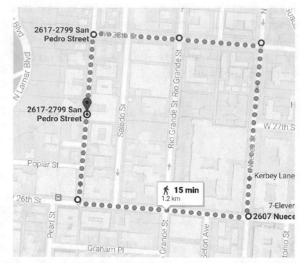

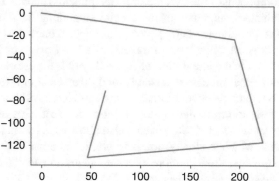

conservation law, with triangular flux function. The model is equivalently written as an *Hamilton Jacobi* equation from which the LWR model is derived [9, 10]. Using this decomposition, we write the problem of estimating traffic density on a section of road as a mixed integer linear program (MILP) [5]. This formulation can also be extended to networks, as in [24]. The solution to the MILP corresponds to a vector of current traffic densities, which can be interpreted as a traffic density map.

The LWR model [25] is described by the following Partial Differential Equation (PDE):

$$\frac{\partial \mathbf{k}(t, x)}{\partial t} + \frac{\partial \psi \mathbf{k}(t, x)}{\partial x} = 0 \qquad (4.3)$$

encoded by the following Hamilton-Jacobi [10] partial differential equation:

$$\frac{\partial \mathbf{M}(t, x)}{\partial t} - \psi \left(-\frac{\partial \mathbf{M}(t, x)}{\partial x} \right) = 0 \qquad (4.4)$$

The function $\psi(\cdot)$ defined in Eq. (4.4) is the *Hamiltonian*. The B-J/F [4, 14] solutions to Eq. (4.4) are fully characterized by a *Lax-Hopf* formula [2, 9], which was initially derived using the control framework of viability theory [1]. We assume that the Hamiltonian is piecewise affine and continuous [13]:

$$\psi(\rho) = \begin{cases} v_f \rho & : \rho \in [0, k_c] \\ w(\rho - \kappa) & : \rho \in [k_c, \kappa] \end{cases} \tag{4.5}$$

The estimation of traffic requires the knowledge of the demand and supply flows applying to the boundaries of the transportation network, and of the splitting coefficients at each intersection. The splitting coefficients can be estimated over time in each cluster, by computing the average flow ratio at each intersection of the corresponding subnetwork. The boundary flows are inputs to each traffic estimation subproblem, and are determined iteratively by a consensus-type algorithm.

4.6.1.1 Input Data

On each segment of the road network, the input data can be written as an affine initial, boundary or internal condition (including internal density condition), as follows.

Affine Initial, Upstream/Downstream Boundary and Internal Conditions Let us define $\mathbb{K} = \{0, \ldots, k_{max}\}$, $\mathbb{N} = \{0, \ldots, n_{max}\}$, $\mathbb{M} = \{0, \ldots, m_{max}\}$ and $\mathbb{U} = \{0, \ldots, u_{max}\}$. For all $k \in \mathbb{K}$, $n \in \mathbb{N}$, $m \in \mathbb{M}$ and $u \in \mathbb{U}$, we define the following functions, respectively called initial, upstream, downstream internal flow and internal density conditions:

$$M_k(t, x) = \begin{cases} -\sum_{i=0}^{k-1} \rho_{ini}(i)X \\ -\rho_{ini}(k)(x - kX) & \text{if } t = 0 \\ & \text{and } x \in [kX, (k+1)X] \\ +\infty & \text{otherwise} \end{cases} \tag{4.6}$$

$$\gamma_n(t, x) = \begin{cases} \sum_{i=0}^{n-1} q_{in}(i)T \\ +q_{in}(n)(t - nT) & \text{if } x = \xi \\ & \text{and } t \in [nT, (n+1)T] \\ +\infty & \text{otherwise} \end{cases} \tag{4.7}$$

$$\beta_n(t, x) = \begin{cases} \sum_{i=0}^{n-1} q_{out}(i)T \\ +q_{out}(n)(t - nT) \\ -\sum_{k=0}^{k_{max}} \rho(k)X & \text{if } x = \chi \\ & \text{and } t \in [nT, (n+1)T] \\ +\infty & \text{otherwise} \end{cases} \tag{4.8}$$

$$\mu_m(t,x) = \begin{cases} L(m) + r(m)(t - t_{\min}(m)) & \text{if } x = x_{\min}(m) \\ \quad + v^{\text{meas}}(m)(t - t_{\min}(m)) \\ \quad \text{and } t \in [t_{\min}(m), t_{\max}(m)] \\ +\infty & \text{otherwise} \end{cases} \tag{4.9}$$

$$\Upsilon_u(t,x) = \begin{cases} L(u) - \rho(u)(x - x_{\min_\rho}(u)) & \text{if } x \in [x_{\min_\rho}(u), x_{\max_\rho}(u)] \\ \quad \text{and } t = t_\rho(u) \\ +\infty & \text{otherwise} \end{cases} \tag{4.10}$$

where $v^{\text{meas}}(m) = \frac{x_{\max}(m) - x_{\min}(m)}{t_{\max}(m) - t_{\min}(m)}$.

In the above definition, internal density conditions (4.10) are specific to model density sensors that are inside the computational domain, generating data at positive times. The data generated by density sensors at time zero would correspond to an initial density. Flow sensors generate upstream (respectively downstream) boundary conditions when located at the upstream (respectively downstream) boundary of the computational domain, and internal conditions associated with zero velocity when located inside the computational domain. Note that the affine initial, upstream/downstream boundary and internal conditions defined above for the HJ PDE (4.4) are equivalent to constant initial, upstream/downstream boundary and internal conditions for the LWR PDE (4.3). The domain of these conditions is illustrated in Fig. 4.7.

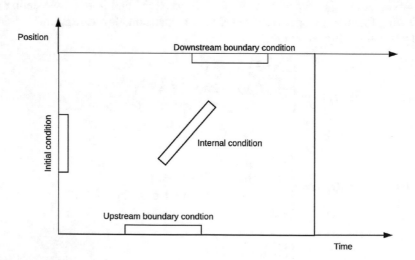

Fig. 4.7 Illustration of the domains of definitions of the different block boundary conditions considered in this article

4.6.2   Traffic State Estimation Using Mixed Integer Linear Programming

We consider a set of block boundary conditions as in Fig. 4.7, generated by measurements originating from fixed sensors (initial, upstream, downstream, and internal flow/density conditions) or probe vehicles (internal velocity conditions). Some of the coefficients of these block boundary conditions are known, while some others need to be estimated. For example, a vehicle generating an internal velocity condition block (4.9) will generate measurements of $x_{\min}(m)$, $x_{\max}(m)$, $t_{\min}(m)$, $t_{\max}(m)$, and $v^{\mathrm{meas}}(m)$. The remaining coefficients $L(m)$ and $r(m)$, corresponding to the vehicle label and passing rate, are unmeasured. Let us call V the vector space of all unknown coefficients, associated with all block boundary conditions of the traffic estimation problem. The possible values of these coefficients are restricted by the physics of the problem, which we refer to as model constraints [6]. Similarly, the measurement data generated by the sensors constrains the possible values of these coefficients. These constraints are similarly called *data constraints*. An important and nontrivial result of [11] is that all these constraints are explicit. The extensive list of all constraints can be found in [7], which we do not write explicitly for compactness. The main result is the following:

Mixed Integer Linear Inequality Property The model constraints are mixed integer linear in the unknown coefficients v of the boundary condition blocks. The data constraints [11] are linear in the unknown coefficients v of the boundary condition blocks. The constraints encoded by merge or diverge junctions [16], or one to one junctions (resulting, for example, from a change of number of lanes or speed limit on a link) are mixed integer linear in the unknown coefficients v of the boundary condition blocks.

Hence, the set of possible traffic scenarios compatible with the data and the model can be written as $\{y \mid Ay \leq b$ and $Cy \leq d\}$, where y represents the decision variable of the problem, which consists of the unknown boundary coefficients v and additional Boolean variables required to write the model constraints as mixed integer linear. To select a solution among all possible choices, we optimize a function of y, which can, for example, be the L_1 norm of y, if the data is sparse. This objective function yields estimates that have the least amount of features (for example, flow and density changes), and are particularly adapted to traffic monitoring situations in which data is sparse. For general linear objective functions, the problem of estimating the state of traffic on the subnetwork can thus be written as:

$$\text{Min. } c^T y$$
$$\text{s. t. } \begin{cases} Ay \leq b \\ Cy \leq d \end{cases} \tag{4.11}$$

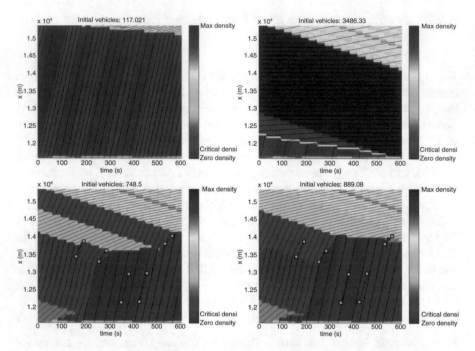

Fig. 4.8 Traffic state estimation example. In this figure, we consider a single stretch of highway. The upper two figures consist in state estimates with boundary data only, in the form of measured upstream and downstream boundary flows. The upper left subfigure represents solution to (4.11) minimizing the total accumulation of vehicles at the initial time, while the upper right subfigure maximizes the same objective. The lower two figures represent the solutions to the same problem, in which five additional internal conditions (representing probe vehicle measurements) are added

The above estimation process is illustrated for two objective functions in Fig. 4.8, over data generated over a single link. An example traffic state estimation problem over a network is illustrated in Fig. 4.8. The later article contains the detailed theory associated with the estimation framework, which is out of the scope of the present article.

4.6.3 Boundary Conditions Estimation

One of the difficulties associated with estimating the state of traffic flow over contiguous subnetworks is enforcing the conservation of flows at the boundaries of each subnetwork. Since each estimation problem is solved by each cluster independently, the boundary flows will not be consistent in general. For example, if a link a is split between subnetworks i and j, the outgoing flow from a, denoted as $q_{i,\text{out}}(\cdot)$ and the incoming flow from a, $q_{j,\text{in}}(\cdot)$ have to be equal. However, the function $q_{i,\text{out}}(\cdot)$ results from the solution associated with subnetwork i, and $q_{j,\text{in}}(\cdot)$ results from the solution associated with subnetwork j, which will not match in general.

Enforcing these conservation equations at the boundaries of each subnetwork requires some communication between subnetworks, and cannot be solved easily, since the boundary conditions associated with a hyperbolic conservation law (such as the LWR model) are *weak*, in that they do not necessarily strongly apply, as shown in [33].

4.6.3.1   Consensus Filtering

Let us consider a cluster $c_i \in \mathscr{I}$, with $\mathscr{S}(i)$ the set of neighboring clusters c_j, defined as the set of clusters c_j for which the subnetworks c_i and c_j have at least one vertex in common $v_{i,j}$. Let us further assume[3] that each common vertex $v_{i,j}^k$ is only connected to two links, $L_{i,j}^k(i)$ and $L_{i,j}^k(j)$ within clusters i and j, respectively.

Let us further distinguish incoming and outgoing links in cluster i as $I_{i,j}^k(i)$ and $O_{i,j}^k(i)$, and denote as $\mathscr{I}(i, j)$ and $\mathscr{O}(i, j)$ the set of corresponding indices.

Let c_i, A_i and b_i correspond to the objective and constraints of each subnetwork i estimation problem, and let y_i correspond to the optimal solutions of each subproblem. The variable y_i contains the optimal boundary flows $q_{L_{i,j}^k(i)}(p)_{p \in \{1,\dots,n\}}$ occurring at each link $L_{i,j}^k(i)$, corresponding to $k \cdot n$ variables, where n represents the number of time steps on which the estimation is run. The solution y_j to the traffic estimation problem corresponding to subnetwork j similarly contains the optimal boundary flows $q_{L_{i,j}^k(j)}(p)_{p \in \{1,\dots,n\}}$ occurring at each link $L_{i,j}^k(j)$. Since these problems are solved independently, the boundary flows are not identical in general.

To regularize the solutions, we propose the following iterative scheme:

1. Solve for y_i and y_j, $j \in \mathscr{S}(i)$.
2. Compute average flows $r_{L_{i,j}^k} = \dfrac{q_{L_{i,j}^k(i)}(p) + q_{L_{i,j}^k(j)}(p)}{2}$, $p \in \{1,\dots,n\}$ between adjacent clusters.
3. Solve the original problem with an additional objective term minimizing the difference between boundary flows and the corresponding previously computed average flows.
4. Return.

---

[3]This assumption is not restrictive in practice, since the definition of the boundaries of each cluster can be adjusted to meet this requirement.

This algorithm in effect regularizes the flows between each cluster, to minimize the flow discrepancies between each cluster. Note that a global optimization formulation of the traffic state estimation problem over the entire network could be considered based upon the previously introduced framework. This global optimization framework would ensure that flows between clusters agree, at the expense of privacy, since it would require all traffic data to be sent to a centralized server. An example distributed traffic state estimation following the process described above is shown in Fig. 4.9.

4.7   Conclusion

This chapter presents a novel traffic monitoring architecture based on a combination of fixed `Bluetooth` or `WiFi` readers in conjunction with Inertial Measurement Unit (IMU) based probe vehicles. This system does not rely on classical positioning devices such as the GPS, making it immune to GPS perturbations due to multipath effects or GPS spoofing. The immediate benefits of such a system are its very low marginal cost (particularly when a reader infrastructure is already deployed) and its privacy by design characteristics. Since the estimation process is decentralized, an attacker compromising part of the system only has access to partial information about the users in the corresponding area, making this system very robust to attacks. We show that the data generated by IMUs allow the extraction of relevant traffic measurement data, which cannot be achieved using classical positioning systems due to their measurement uncertainty. We also show that the state estimation can be performed on the reader nodes using any type of state estimation algorithm (for example, based on Kalman Filtering, or on an optimization formulation of the estimation problem to be solved). Each subproblem is associated with unknown boundary conditions at the edge of the corresponding subnetwork. To make these boundary conditions agree, a consensus-based approach is proposed. Future work will deal with the extension of this system to other type of measurements related to urban, arterial, or highway operations, such as the estimation of pavement condition, the detection of the presence of disabled vehicles from turning and acceleration patterns, and the estimation of dangerous hotspot locations where accidents are likely, based on the estimated inputs of the drivers. These inputs can be estimated from inertial measurements, using a dynamical model of the vehicle. This work could also be extended to DSRC-based systems, in which RSEs (Roadside Equipment) would play the role of the readers, and would process local traffic data generated by DSRC-equipped vehicles.

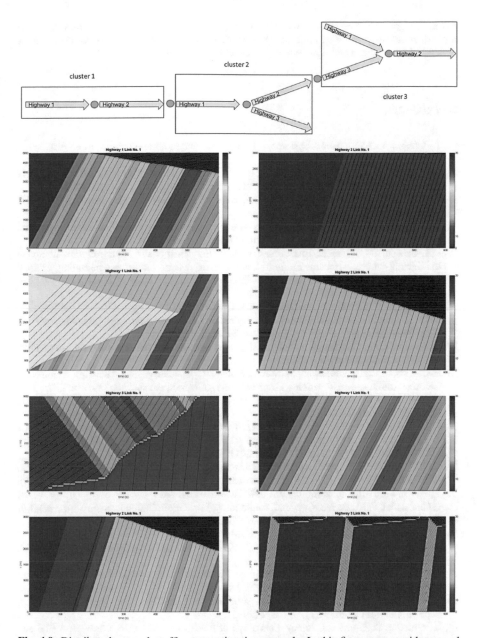

Fig. 4.9 Distributed network traffic state estimation example. In this figure, we consider a road network partitioned into three sub-clusters (top subfigure). The bottom subfigures correspond to the state estimates in all links, obtained by iteratively computing the computation of independent network state estimation problems (for each cluster), and regularizing the boundary conditions between each subnetwork. The upper two subfigures correspond to cluster 1, and the lower six subfigures correspond to the traffic state on clusters 2 and 3

Acknowledgements The authors would like to thank the Texas Department of Transportation for supporting this research under project 0-6838, Bringing Smart Transport to Texans: Ensuring the Benefits of a Connected and Autonomous Transport System in Texas.

References

1. J.-P. Aubin, *Viability Theory*. Systems and Control: Foundations and Applications (Birkhäuser, Boston, 1991)
2. J.-P. Aubin, A.M. Bayen, P. Saint-Pierre, Dirichlet problems for some Hamilton-Jacobi equations with inequality constraints. SIAM J. Control Optim. **47**(5), 2348–2380 (2008)
3. S. Bae, A. Kwasinski, Spatial and temporal model of electric vehicle charging demand. IEEE Trans. Smart Grid **3**(1), 394–403 (2012)
4. E.N. Barron, R. Jensen, Semicontinuous viscosity solutions for Hamilton-Jacobi equations with convex Hamiltonians. Commun. Partial Differ. Equ. **15**, 1713–1742 (1990)
5. E.S. Canepa, C.G. Claudel, Exact solutions to traffic density estimation problems involving the lighthill-whitham-richards traffic flow model using mixed integer programming, in *2012 15th International IEEE Conference on Intelligent Transportation Systems (ITSC)* (IEEE, Piscataway, 2012), pp. 832–839
6. E.S. Canepa, C.G. Claudel, A framework for privacy and security analysis of probe-based traffic information systems, in *Proceedings of the 2nd ACM International Conference on High Confidence Networked Systems* (ACM, New York, 2013), pp. 25–32
7. E. Canepa, C.G. Claudel, A model-based framework for user privacy analysis using probe traffic data. Technical report (2017)
8. E. Canepa, E. Odat, A. Dehwah, M. Mousa, J. Jiang, C. Claudel, A sensor network architecture for urban traffic state estimation with mixed eulerian/lagrangian sensing based on distributed computing, in *International Conference on Architecture of Computing Systems* (Springer, Berlin, 2014), pp. 147–158
9. C.G. Claudel, A.M. Bayen, Lax-Hopf based incorporation of internal boundary conditions into Hamilton-Jacobi equation. Part I: theory. IEEE Trans. Autom. Control **55**(5), 1142–1157 (2010). https://doi.org/10.1109/TAC.2010.2041976
10. C.G. Claudel, A.M. Bayen, Lax-Hopf based incorporation of internal boundary conditions into Hamilton-Jacobi equation. Part II: computational methods. IEEE Trans. Autom. Control **55**(5), 1158–1174 (2010). https://doi.org/10.1109/TAC.2010.2045439
11. C.G. Claudel, A.M. Bayen, Convex formulations of data assimilation problems for a class of Hamilton-Jacobi equations. SIAM J. Control Optim. **49**, 383–402 (2011)
12. J. Cortés, G.E. Dullerud, S. Han, J. Le Ny, S. Mitra, G.J. Pappas, Differential privacy in control and network systems, in *2016 IEEE 55th Conference on Decision and Control (CDC)* (IEEE, Piscataway, 2016), pp. 4252–4272
13. C.F. Daganzo, A variational formulation of kinematic waves: basic theory and complex boundary conditions. Transp. Res B **39B**(2), 187–196 (2005)
14. H. Frankowska, Lower semicontinuous solutions of Hamilton-Jacobi-Bellman equations. SIAM J. Control Optim. **31**(1), 257–272 (1993)
15. M. Gruteser, D. Grunwald, Anonymous usage of location-based services through spatial and temporal cloaking, in *Proceedings of the 1st International Conference on Mobile Systems, Applications and Services* (ACM, New York, 2003), pp. 31–42
16. K. Han, T.L. Friesz, T. Yao, A link-based mixed integer LP approach for adaptive traffic signal control (2012, Preprint). arXiv:1211.4625
17. B. Hoh, M. Gruteser, R. Herring, J. Ban, D. Work, J.C. Herrera, A.M. Bayen, M. Annavaram, Q. Jacobson, Virtual trip lines for distributed privacy-preserving traffic monitoring. *Proceedings of the 6th International Conference on Mobile systems, Applications, and Services, MobiSys 2008*, Breckenridge (2008), pp. 15–28

18. Z. Hou, J.-X. Xu, J. Yan, An iterative learning approach for density control of freeway traffic flow via ramp metering. Transp. Res. Part C Emerg. Technol. **16**(1), 71–97 (2008)
19. J. Krumm, Inference attacks on location tracks, in *Pervasive Computing* (Springer, Berlin, 2007), pp. 127–143
20. J. Krumm, A survey of computational location privacy. Pers. Ubiquit. Comput. **13**(6), 391–399 (2009)
21. C. Lee, B, Hellinga, F. Saccomanno, Evaluation of variable speed limits to improve traffic safety. Transp. Res. Part C Emerg. Technol. **14**(3), 213–228 (2006)
22. E.J. Lefferts, F.L. Markley, M.D. Shuster, Kalman filtering for spacecraft attitude estimation. J. Guid. Control Dyn. **5**(5), 417–429 (1982)
23. Y. Li, E. Canepa, C. Claudel, Efficient robust control of first order scalar conservation laws using semi-analytical solutions. Discrete Contin. Dyn. Syst. Ser. S **7**(3), 525–542 (2014)
24. Y. Li, E. Canepa, C. Claudel, Optimal control of scalar conservation laws using linear/quadratic programming: application to transportation networks. IEEE Trans. Control Netw. Syst. **1**(1):28–39 (2014)
25. M.J. Lighthill, G.B. Whitham, On kinematic waves. II. A theory of traffic flow on long crowded roads. Proc. R. Soc. Lond. **229**(1178), 317–345 (1956)
26. X.-Y. Lu, P. Varaiya, R. Horowitz, D. Su, S. Shladover, Novel freeway traffic control with variable speed limit and coordinated ramp metering. Transp. Res. Rec. J. Transp. Res. Board **2229**, 55–65 (2011)
27. M. Mousa, M. Abdulaal, S. Boyles, C. Claudel, Inertial measurement unit-based traffic monitoring using short range wireless sensor networks, in *Transportation Research Board 94th Annual Meeting*, number 15-4072 (2015)
28. M. Mousa, K. Sharma, C. Claudel, Automatic calibration of device attitude in inertial measurement unit based traffic probe vehicles, in *2016 15th ACM/IEEE International Conference on Information Processing in Sensor Networks (IPSN)* (IEEE, Piscataway, 2016), pp. 1–2
29. M. Papageorgiou, H. Hadj-Salem, J.-M. Blosseville, Alinea: a local feedback control law for on-ramp metering. Transp. Res. Rec. **1320**(1), 58–67 (1991)
30. S.T. Peddinti, N. Saxena, On the limitations of query obfuscation techniques for location privacy, in *Proceedings of the 13th International Conference on Ubiquitous Computing* (ACM, New York, 2011), pp. 187–196
31. D. Pisarski, C. Canudas-de Wit, Nash game-based distributed control design for balancing traffic density over freeway networks. IEEE Trans. Control Netw. Syst. **3**(2), 149–161 (2016)
32. W. Premerlani, P. Bizard, Direction cosine matrix IMU: theory. DIY Drone: USA, pp. 13–15, 2009
33. I.S. Strub, A.M. Bayen, Weak formulation of boundary conditions for scalar conservation laws. Int. J. Robust Nonlinear Control **16**(16), 733–748 (2006)
34. Y. Sun, D.B. Work, Scaling the Kalman filter for large-scale traffic estimation. IEEE Trans. Control Netw. Syst. (2017). http://ieeexplore.ieee.org/document/7855709/; http://dx.doi.org/10.1109/TCNS.2017.2668898
35. R. Wang, D.B. Work, R. Sowers, Multiple model particle filter for traffic estimation and incident detection. IEEE Trans. Intell. Transp. Syst. **17**(12), 3461–3470 (2016)
36. D. Work, S. Blandin, O. Tossavainen, B. Piccoli, A. Bayen, A distributed highway velocity model for traffic state reconstruction. Appl. Res. Math. eXpress (ARMX) **1**, 1–35 (2010)
37. X. Wu, S. Subramanian, R. Guha, R.G. White, J. Li, K.W. Lu, A. Bucceri, T. Zhang, Vehicular communications using DSRC: challenges, enhancements, and evolution. IEEE J. Sel. Areas Commun. **31**(9), 399–408 (2013)

Chapter 5
Data, Methods, and Applications of Traffic Source Prediction

Chengcheng Wang and Pu Wang

5.1 Introduction

The rapid urbanization occurring globally has caused an imbalance between the fast-rising demand for transportation and the limited land available for transportation infrastructure. Therefore, traffic congestion is ubiquitous in many cities, and solving or mitigating such congestion is crucial for transportation efficiency, energy conservation, environmental protection, and human health. In recent years, much research and engineering practices have focused on the analysis, avoidance, and mitigation of traffic congestion. A common strategy has been to isolate the locations closely linked to traffic congestion in order to implement targeted traffic management to more effectively mitigate congestion. Traditional traffic surveys only record samples of the origin and destination of trips, lacking information on how traffic congestion is caused. To locate the source of traffic congestion, real-time and high-resolution transportation data is needed, yet traffic information recorded by traditional surveys is always inaccurate and easily out of date [1]. Fortunately, the fast development of sensing and computing techniques in recent years has provided the necessary data for researchers to pinpoint traffic-congested driver sources and develop novel traffic management strategies.

The concept of driver sources was first proposed by Wang et al. [1], where large-scale mobile phone data were used to explore road usage patterns. The exciting finding from Wang et al. [1] was that the major driver sources for most road segments are limited. Afterwards, dynamic driver sources, which can be better applied in real-time traffic control, were proposed by Wang et al. [2].

C. Wang · P. Wang (✉)
School of Traffic and Transportation Engineering, Central South University, Changsha, Hunan, China
e-mail: wangpu@csu.edu.cn

© Springer International Publishing AG, part of Springer Nature 2019
S. V. Ukkusuri, C. Yang (eds.), *Transportation Analytics in the Era of Big Data*, Complex Networks and Dynamic Systems 4, https://doi.org/10.1007/978-3-319-75862-6_5

With the availability of more high-resolution traffic data, such as radio-frequency identification (RFID), dynamic driver sources can be located at higher temporal and spatial resolutions. The concept of driver source can be modified to passenger source, which was applied in the passenger flow analysis of urban rail transit (URT) networks [3–5]. Traffic source is a more general term that includes both the sources of drivers in road networks and the sources of passengers in public transportation networks. A number of applications have been developed to mitigate traffic congestion based on traffic source information, and the mitigation methods can basically be grouped into three classes: traffic demand restrictions [1], traffic routing guidance [4, 6, 7], and transportation infrastructure upgrades [2]. In the following, we elaborate on the data used, methods developed, and applications proposed to predict traffic sources.

5.2 Data

Traffic demand is fundamental information for traffic planning and transportation management. In today's Big Data era, various types of data that capture abundant traffic and transportation information have become increasingly available. These transportation data include data recorded by mobile phones, RFID chips, video cameras, etc. In this section, we first review the different traffic data sources that are used to locate traffic sources that contribute the major volumes of flows on congested roads.

5.2.1 Mobile Phone Data

Because of the timeliness' absence of traffic origin–destination (OD) matrices, it is difficult to obtain meaningful data from traditional traffic surveys, which are always expensive. In recent years, with massive expansion in mobile phone use, human mobility information can be easily, efficiently, and cheaply collected. Mobile phone towers are located and densely distributed in urban areas and thus can provide detailed information on daily human mobility. Human spatiotemporal information, including detailed time-variant travel demand data, can be extracted from large-scale mobile phone data, providing foundations for traffic planning. Mobile phone data are widely available in large cities because they are originally collected by the billing process. This means that an approach based on mobile phone data can be easily extended to multiple cities.

Mobile phone data provide a source of generating data on the distribution of travel demands on an unprecedented scale. Indeed, the wide emergence of mobile phone data has stimulated rapid developments in human mobility. Gonzalez et al. [8] analyzed the trajectories of 100,000 mobile phone users through their billing records and uncovered several universal human mobility laws. Based on mobile phone data,

Song et al. [9] discovered that human beings are highly predictable regardless of travel distance, age, and sex, thus establishing the theoretical foundation for developing accurate predictive models of human mobility. Devillea et al. [10] used more than 1 billion mobile phone call records from Portugal and France to estimate dynamic population densities on a national scale. With human mobility models, human movements can be predicted and, consequently, OD matrices estimated.

Mobile phone call detail records (CDRs) comprise the most general type of mobile phone data. However, there are two limitations of CDR. First, they contain sparse and irregular records, in which user displacements (consecutive nonidentical locations) are often observed with long travel intervals. Second, tower shifting exists in the data, which represents no actual displacements but is instead caused by the operator often balancing call traffic among adjacent towers. In summary, CDR data are more suitable for obtaining the statistical distribution of travel demands. The other type of mobile phone data is mobile phone signaling data, in which user locations are recorded at regular time intervals (e.g., 30 min). Mobile phone signaling data are much better for estimating travel demand information; however, given their heavy collecting load, such data are few and usually not recorded for a long period.

5.2.2 Radio-Frequency Identification Data

Travel information extracted by mobile phone data is greatly improved in terms of timeliness and practicability but has shortcomings in terms of accuracy and real-time performance. RFID data, which records vehicular locations through RFID equipment, could better overcome these deficiencies and serve as a kind of high-resolution data. RFID sensing equipment can accurately record the exact time and location when an RFID-equipped vehicle passes [11]. Densely distributed RFID sensing equipment can record information at a high frequency. Usually, the vehicle's ID, the ID of the RFID station the vehicle passes, and the exact time the vehicle passed the station are recorded, so the RFID-equipped vehicles can be traced and the required traffic information collected.

With the advantage of high resolution, the recording of vehicles' spatiotemporal information is more accurate and captured with relatively high frequencies, and the RFID data are more useful for generating dynamic travel demands and developing real-time traffic management systems. Khateeb et al. [11] proposed dynamic traffic light sequence algorithm using RFID, which could improve the efficiency of traffic management. Yang [12] employed Complex Event Processing technology based on RFID to improve real-time event detection, while Wen et al. [13] presented an intelligent traffic management expert system using RFID technology that provided traffic data collection and control information. However, there is one current limitation of RFID data, namely that RFID devices are not installed in all vehicles, which means that only some of the vehicles passing an RFID sensing station can be detected and recorded. With the installation of RFID devices in more vehicles, traffic demand information could be obtained on a larger scale.

5.2.3   Subway Card Data

Subways comprise one of the most important modes of human travel, and therefore, subway card data are important in passenger flow analysis. In a subway system, each time a passenger uses his/her card when entering the subway, time, card ID, subway line ID, subway station ID, and fare collection device ID are recorded. An advantage of subway card data is that when a passenger exits the subway, the station is recorded, which provides concrete information on passenger destinations. Because of the popularity of subway cards, the approach of traffic source prediction based on subway card data can be easily extended to cities with subway systems.

Subway card data offer complete access to station information, the data defects and irregularities are few, the extraction of OD information is more accurate, and the results are usually more credible. Using precise subway traffic information, passenger flow prediction and passengers' demand pattern have been investigated in depth in recent years. Wei and Chen [14] developed a hybrid forecasting approach combining empirical mode decomposition and back-propagation neural networks; this approach could perform well and stably in forecasting short-term passenger flow. Sun et al. [15] used smartcard data to investigate and understand the demand pattern of passengers and extract a train's trajectory; this could help transit agencies better prepare to respond to failures more timely and effectively.

5.2.4   Other Data and Characteristics of Different Data Sources

Different data sources have different characteristics and application scopes in congestion source prediction. Mobile phone data has advantages of long recording time, wide spatial coverage, and low collection cost, but it is poor at data accuracy and frequency. RFID is characterized with two important advantageous features: the high spatial-temporal resolution and vehicles' identities being recorded. Therefore, using RFID data can locate dynamic driver sources. However, RFID data is limited with its penetration rate and spatial coverage. Conversely, subway card data is a good data source that provides accurate OD information for locating passenger sources in subway networks.

Besides the three kinds of data introduced above, video camera data, which is similar to RFID data, can also be used to locate driver sources. The differences between video camera data and RFID data is the way vehicle information is obtained: one is through image processing, the other is through RFID equipment. Another data source is GPS data, which records vehicle's location every several seconds and is characterized with a higher data recording frequency; however, GPS data cannot provide the full OD information.

5.3 Methods

Using different kinds of traffic data, traffic demand can be estimated. There are different methods of estimating traffic demand based on different data characteristics. To explore road usage patterns, traffic demand is assigned to a road segment as close to actual traffic as far as possible. Through the exploration of traffic flow, the major driver sources of each road segment are considered. The methods for locating driver sources, dynamic driver sources, and passenger sources are summarized in the following.

5.3.1 Travel Demand Estimation

Travel demand is an important decision support for traffic planning, management, and control, and there are different methods of estimating it using different types of data (Fig. 5.1).

Regarding mobile phone data, the most challenging aspects of analysis are sparse and irregular records. In Wang et al. [1], to more accurately extract users' travel demands and at the same time ensure that enough travel demand information was extracted, a trip was usually defined as a displacement occurring within 1 h, and therefore the travel time window was set to 1 h. Because location information is lost when users do not use phones, the transient origin destination (t-OD) matrix was put forward. Then, travel demands could be generated independent of the frequency of phone activity by selecting stable phone user groups. To offset the deviation caused by the irregular distribution of mobile phone users, the ratio of the population to the number of mobile phone users in a zone was applied in Wang et al. [1] to scale up or scale down the number of trips. Furthermore, because of the different transportation modes of trips, the vehicle using rate was calculated to determine the number of trips in each zone traveled by vehicles. Finally, the estimated t-OD was defined based on vehicle trips and the entire population. Because mobile phone data cannot record detailed geographical information, only the zone-based t-OD can be obtained using the above measures. To facilitate the traffic assignment, one intersection within the zone was randomly selected in Wang et al. [1] to convert the zone-based t-OD to the intersection-based t-OD. In cases when no intersection was found in the zone, the origin or destination in the intersection-based t-OD was determined by randomly selecting one intersection in the nearest-neighboring zone.

Some research has focused on travel demand estimation based on mobile phone data. In Iqbal et al. [16], for example, a methodology similar to that described above to develop OD matrices using mobile phone CDRs was proposed. In this methodology, the time-stamped Base Transceiver Station tower locations of each user are first extracted from CDR data. Then, tower-to-tower transient OD matrix was developed by them. The tower-to-tower t-OD must be converted to node-to-node t-OD before applying them; this is accomplished by connecting origin and

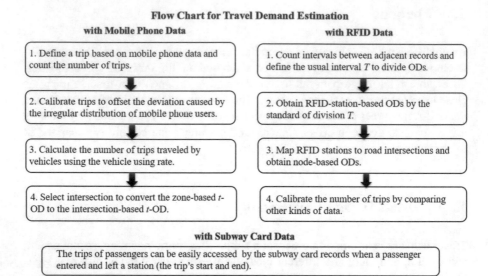

Fig. 5.1 Flow chart of travel demand estimation with different kinds of transportation data

destination towers to corresponding nodes of the network. Moreover, after obtained node-to-node t-OD matrix, simulated traffic flow was generated in the road network. Combining simulated traffic flow and actual traffic flow (generating from video data), this node-to-node t-OD matrix could be scaled up to developed the actual OD matrix. In Gong et al. [17], a method for estimating OD distribution among urban mega traffic analysis zones based on mobile phone data was proposed. Based on this method, the influence of travel information loss is reduced.

RFID data can record the time and location of vehicles trips more accurately compared to mobile phone data, which is more helpful in obtaining accurate OD matrices. Tracing the RFID records of each vehicle, it is easy to find that most time intervals between two adjacent records are distributed in a small range of T. Therefore, the time interval T could be assumed to be the standard for trip separation. For most time intervals smaller than T, the records are assumed to belong to the same trip. For a few consecutive records in which the time interval is larger than the usual T, we usually consider the records as belonging to different trips. Using this method, we can obtain RFID-station-based ODs. In order to facilitate assignment of the trips to segments, each RFID station should be mapped to a road intersection, and node-based ODs are thus obtained. However, since RFID devices have not yet been fully popularized, we usually need to scale up the number of trips between each pair of sources and destinations by comparing them to other kinds of data.

Using daily subway card records, which provide detailed information regarding when a passenger entered and left a station (i.e., the trip's start and end), the daily trips of all passengers can be easily accessed. Sometimes, however, OD matrices for different peak hours are usually needed, e.g., a case of morning peak OD matrix is described in He et al. [4]. The trips that start at the first station after 8 a.m. and at the last station before 9 a.m. belong, of course, to the morning peak OD matrix. In He et al. [4], for the trips that start and end around the morning peak hour, the first station after 8 a.m. and the last station before 9 a.m. for each trip was located by calculating the shortest path of each trip. These stations were used as the new origin and destination of the trip to generate the morning peak hourly OD matrix.

5.3.2 Traffic Flow Assignment

After travel demand has been estimated, different methods of assigning trips to road networks can be utilized. The Dijkstra algorithm [18] is the most classical method of assigning trips to road networks. The algorithm is used to solve the shortest-path problem for a graph. In traffic field, travel time of each segment is usually used as edge's weight. Furthermore, it is the most basic of traffic flow assignment algorithms. In "all-or-nothing" traffic assignment, all trips are assigned to the road network by selecting the shortest path of each trip. Then, the traffic flow on each road segment can be estimated. It is a static nonequilibrium model approach in which the same trip is assigned to the same shortest path and one that only considers free traffic conditions without the additional travel time incurred due to traffic congestion. "All-or-nothing" traffic assignment is usually used in subway networks or simple road networks for which there is no need to consider additional travel costs in general and can also be used in road networks with few trips and where congestion does not usually occur.

All-or-nothing traffic assignment ignores the dynamical change of road segment travel time that occurs because of the cost of congestion. The incremental traffic assignment (ITA) [19] method could solve this problem by updating travel costs that consider travel congestion. In the ITA method, original trips are usually first split into different sub-trips that contain different percentages of the original trips. First, one of sub-trips is assigned to the network using the free travel time. After that, the actual travel time in a road segment is updated using the Bureau of Public Roads function. Then, the next sub-trips are assigned using the actual travel time, and the process is continued until all sub-trips are assigned to the road network. ITA is an improvement of all-or-nothing assignment methods, and because of its convenience and the fact that it considers additional travel costs, it is often used in complex road networks with a large number of trips [1, 2].

The ITA method incorporates the change of travel time, but in the actual traffic conditions each driver always uses the route to minimize his or her travel cost, which attains user equilibrium (UE). There are many methods of computing UE, including the method of successive average algorithm, which is a simple, classical, and efficient assignment algorithm that approximately achieves the UE state. This algorithm assigns all trips to segments and upgrades travel times constantly until the traffic flow reaches the UE state. Another convenient algorithm is Frank–Wolfe

[20] algorithm, and the system approaches the UE state more quickly. These two algorithms are both easy to realize but are slow in attaining the optimal UE state because of constant circulation. Because these kinds of algorithms consider user benefits, they most closely approximate the actual traffic situation and thus are often used in the analysis of traffic conditions. However, due to their high time cost, especially when applied to complex road networks, a more convenient algorithm is sometimes chosen to replace them.

5.3.3   Locating the Driver Sources

Using an appropriate traffic assignment method from those described above, each trip is assigned to the road network to provide estimated traffic flows. The ratio of traffic flow to the capacity of a road is called volume over capacity (VOC), and it can reflect the extent of traffic congestion in some ways. We usually recognize that a road is congested when the traffic flow is greater than its available capacity. From the investigation of road segment usage reported in Wang et al. [1], mobile phone data in the San Francisco Bay Area (California, USA) and Boston (Massachusetts, USA) were used to analyze traffic flow, betweenness centrality, and VOC in the two urban areas. Traffic flow follows mixed exponential distribution in the two urban areas, and the arterial roads' and highways' betweenness centrality can be separately approximated by the power-law distribution and the exponential distribution. Although topologies of road networks are different, the traffic flow distribution is similar, which indicates an inherent mechanism in road usage patterns. VOC follows an exponential distribution in most road segments. In Wang et al. [1], the surprising finding was that only a few locations caused the congestion. Furthermore, the driver source was defined as a mobile phone user's home location. The process of major driver sources prediction is as follows:

In Fig. 5.2, rectangles represent traffic zones, the red zone H represents the home location of a mobile phone user, the blue zone A and B, respectively, represent the start and end of his or her trip, and the red line represents the predicted route from zone A and zone B. An array variable $S[x]$ (x is the ID of a zone) is defined to quantify the traffic flow contribution from zones to road segments. If a road segment belongs to a part of the predicted route, $S[x = H] + 1$. Then, the traffic flow contribution from each zone to the segment is calculated by counting $S[x]$ from all paths in t-OD matrices. Sorting and analyzing traffic contributions of each zone can reflect the different impact of each zone to the congestion of the road segment. The authors defined the top-ranked sources as those providing the most traffic flow on a road segment as its major driver sources (the major driver sources in total produce 80% of a segment's traffic flow).

Next, the road usage bipartite network is formed to explore the relationship between major driver sources and road segments. In bipartite networks, links only exist in different types of nodes. In the road usage bipartite network, road segments and major driver sources are the two types of nodes, and each road segment is

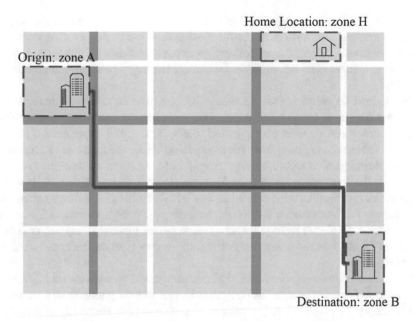

Fig. 5.2 Illustration of a traveler's origin, destination, and home location. The rectangles represent different zones, the green lines and white lines, respectively, represent highways and arterial roads. The red line represents the predicted route from zone A to zone B

connected with its driver sources. In the modeling framework of road usage bipartite network, the degree of a driver source and the degree of a road segment were proposed [1]. The former is the number of road segments for which the driver source is the major driver source, and the latter is the number of major driver sources of the road segment. The driver sources' and road segments' degrees can be approximated by normal distribution and log-normal distribution, respectively. The different distributions of the degree of a driver source and the degree of a road segment reflect different internal relations. First, a similar number of road segments were used by drivers from each driver source. Second, and useful to congestion mitigation, is that only a few driver sources provide the major usage of a road segment.

Static driver sources are useful information for traffic planning and management, but because of their stability they cannot be used for real-time dynamic traffic control. In Wang et al. [2], the dynamical driver sources of a road segment were not defined by a mobile phone user's home location, but rather by the census tracts for the location where the trip started in the time window. In this way, the driver sources information captured the time-variant vehicle sources for a road at different times of day. Similar to the major driver source described above, in Fig. 5.2, the blue zone A represents the start zone of the trip to zone B; if the road segment belongs to the predicted route (red line), $S[x = A] + 1$. After counting $S[x]$ from all paths, the traffic flow contribution from each zone to the road segment is obtained. Finally, a road segment's major dynamical driver sources are defined as the top-ranked sources that, in total, produce 80% of the traffic flow. It is worth noting that the number of major dynamical driver sources follows an exponential distribution, which means that fewer sources must be considered when controlling a road segment's traffic in real time.

For a road segment, t_e (the difference between the actual travel time, t_a, and the free flow travel time, t_f) could also be used measure its level of congestion [1]. If a driver travels through congested roads, he/she will experience a large t_e. If where these drivers come from could be determined, the measures necessary to ease traffic congestion would be easy to realize. In the research described in Wang et al. [1], the total extra travel time T_e generated by driver sources followed an exponential distribution, which indicates that it is feasible to target the small number of congested driver sources that could be defined by the top-ranked T_e of each driver source. Different from segment driver sources, the congested driver sources are more focused on the congestion of the entire city than on the congestion of a segment.

The driver sources based on mobile phone data could provide useful information for traffic control, but because of the sparseness of mobile phone data, its accuracy and instantaneity is defective. The high resolution and high frequency of RFID data greatly overcomes these problems, since each RFID station is considered a driver source, and the major sources of driver and congested driver sources are defined similarly as described before. Because of the higher temporal and spatial resolution of these dynamic driver sources, the RFID-data-based driver source information is more applicable in emergent situations.

Similar to the concept of "driver sources," in the subway network described in He et al. [4], subway stations are defined as "passenger sources." In He et al. [4], OD pairs were first ranked according to their total extra travel cost, and the starting station of each top-ranked OD pair was defined as a congested passenger source and was targeted for releasing routing information. In the URT network described in Wang et al. [3], for a URT segment each passenger's home census tract was defined as a passenger source. Similar to the definition of a "driver source," Wang et al. [3] ranked census tracts by their passenger flow contribution, and defined the top-ranked census tracts that, in total, produced 80% of the passenger flow of each URT segment as the segment's major passenger sources.

5.3.4 Visualization of Driver Sources

After locating the driver sources of a city road network and the passenger sources of the sections between stations, in order to clearly show the available driver source information, an interactive visual analytics system needs to be established. Such a system's workflow consists of two phases: preprocessing and visualization. Construction of the three subsystems—driver source prediction, passenger source prediction, and congested cluster analysis—provides visual charts, particle trajec-

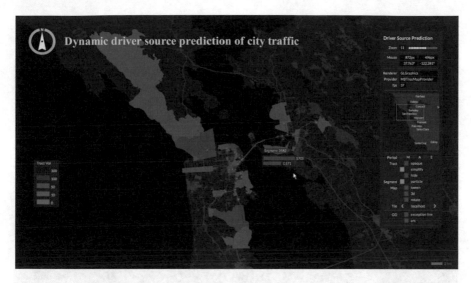

Fig. 5.3 Visualization of dynamic driver sources in the San Francisco Bay area (California, USA) road network

tory views, map zoom, screen shots, custom maps, etc.; meanwhile, the system uses asynchronous loading, offline maps, and vector data compression techniques to provide high-quality and efficient services for users. In consideration of portability, the system supports multiple operating systems. Some driver source information types are shown in Figs. 5.3 and 5.4, which were generated using the interactive visual analytics system described.

5.4   Applications

After discovering that the major traffic flows in congested roads are created by very few driver sources, we began to consider creating new applications based on targeted strategies to mitigate traffic congestion. A straightforward mitigation strategy is to control the traffic demand from targeted driver sources. We can also pinpoint road clusters heavily used by drivers from congested driver sources in order to improve road network efficiency. Moreover, a convenient mitigation strategy involves guiding routes from driver sources.

5.4.1   Traffic Demand Control

After congested driver sources are located, a simple way of mitigating congestion is to reduce the number of trips from congested driver sources. In Wang et al. [1],

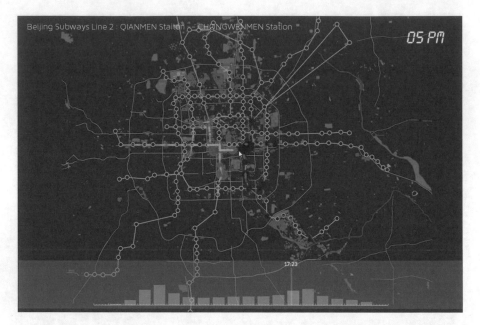

Fig. 5.4 Visualization of passenger sources in the Beijing, China subway network

different percentages of trips from congested driver sources (corresponding to m (0.1–1%) of the total percentage of trips) were reduced. As a reference, the m total percentage of trips were randomly reduced without identifying the congested driver sources. In Wang et al. [1], the total travel time reduction increased linearly with m, and the total travel time reduction achieved using the selective strategy was almost many times more than that achieved using the random strategy, which means that the selective strategy is much more effective in reducing the total additional travel time. From the study of such a high-efficiency selective strategy, the essential reasons for its success, namely that congested road segments are few and traffic flow in those road segments is basically connected to few major driver sources, become obvious.

5.4.2 Pinpointing Road Clusters Heavily Used by Drivers and Infrastructure Upgrades

After defining dynamic congestion driver sources, a connection between a road network and its sources of traffic congestion can be built. In Wang et al. [2], the road cluster was defined as a collection of road segments that were used at least once by congested driver sources, and these roads are closely related to traffic congestion. This finding could be applied to effective mitigation of traffic congestion. One way of mitigating traffic congestion is to increase the capacity of congested roads, which is discouraged because of the likelihood of attracting more vehicles and because of the extra expense. Another way of mitigating traffic congestion is to reduce the number of vehicles on congested roads; this is easy to implement by lowering the speed limits on road segments, as reported in Wang et al. [2]. Both measures were applied in the targeted road clusters, and for comparison, also applied in road clusters that were randomly generated with the same sizes as the targeted road clusters. For both approaches, the selective strategy was much more effective than the random one in reducing total additional travel time and congestion. Comparing these approaches, lowering the speed limit is more appropriate because it is more adaptive and reduces operation/construction costs.

5.4.3   Route Guidance

To mitigate traffic congestion, travelers should be routed to use transportation networks more appropriately and in ways that are more acceptable to them. One routing guidance method is the hybrid routing model. In He et al. [6], a hybrid routing model to alleviate traffic jams was proposed, in which guiding a small portion of travelers to choose the minimum-cost path achieved almost the same congestion mitigation effect as guiding all travelers to choose the minimum-cost path. To investigate the feasibility of hybrid routing in real passenger traffic control, He et al. [7] reported on the use of the hybrid routing model in the Beijing subway network. The results showed that the hybrid routing model can function as an effective and feasible solution for alleviating severe congestion in subway networks since it requires only approximately 20% of passengers to use minimum-cost routing. To improve the impractical measures that inform every traveler of the proper route, an information-releasing framework that only suggests routes from targeted stations was developed and reported in He et al. [4]. The congested passenger sources were publicized, and routing information was released in the targeted stations. The results in He et al. [4] showed that very few stations are targeted for broadcasting, and most stations only provide a small number of suggested routes, which demonstrates the feasibility of the framework. In He et al. [4], the hybrid routing model was also used for the San Francisco road network, and it also proved effective in road traffic improvement.

Another routing guidance method uses information on dynamical driver sources located by RFID data. It is obvious that it is more feasible to route drivers from targeted sources than to route them from all over the urban space. We could consider that only trips from dynamic congested sources are selected for routing guidance. In this routing guidance method, the social good λ is considered (λ is proposed to explore the routing strategies for balancing individual benefit and social good [21]). This routing guidance method can reduce the influence on normal trips by targeting several appointed places, and the dynamic driver source information could reduce the difficulty of real-time traffic control.

5.5 Discussion and Conclusions

To improve the efficiency of transportation systems, there have been many studies on road network topology and urban traffic demand. The studies of road network topology mainly focused on spatial accessibility [22], betweenness centrality [23], and vulnerability [24]. In traffic demand studies, traffic demand estimation [16, 25], a travel demand transportation supply model [26], and travel demand management system [27] were extensively studied. However, most existing studies lack understanding of the internal relation between network and travel demand.

Combined with the internal relation of network analysis and traffic demand, Wang et al. [1] defined a road usage bipartite network and first proposed "driver source" to explore the source of the congestion. From the point view of macroanalysis and real-time control, the "static driver source" can be applied to urban planning while "dynamic driver source" [2] can support real-time traffic control. From the point view of urban congestion and road segment congestion, the "major driver source" is estimated by traffic flow, and the "congested driver source" is calculated by total extra travel time. In Wang et al. [3], "passenger source" was proposed. With the development of machine learning technology, Wang et al. [28] proposed a deep learning method for continuous traffic speed prediction, and designed a novel influence function based on it to recognize congestion sources. Overall, the "traffic source" provided new insight into traffic management and can be applied in targeting some travelers to reduce urban congestion more efficiently.

Similar to driver source, "driver major destinations" have also been analyzed recently. In Gong et al. [29], major providers of traffic information were defined as major destination zones. Gong et al. [29] estimated the spatial distribution of vehicles, and discovered that at different times there are always very few major providers of traffic information for each traffic zone, which suggested that there is no need for information for an entire area. Toole et al. [30] combined numerous algorithms, generated representative OD matrices and route trips, and built an interactive web visualization that showed the trip producing census tracts and attracting census tracts for a road and roads used by the census tract's generated trips. In the URT network, the major passenger sources were not only pinpointed but a URT segment's major passenger destinations were also defined as the top-ranked census tracts that, in total, attracted 80% of the passenger flow of a segment [3]. Based on this finding, a URT segment's vulnerability can be measured by combining the trip failure rate with the number of major passenger sources and major passenger destinations. When the different segments have the same trip failure rate, the segment with the larger number of major passenger sources or destinations is more vulnerable, and it is more difficult to respond to emergencies or accidents. In addition, Wang et al. [5] found that the impacts of the rail transit network were different when passenger flow suddenly increased in different stations. In this paper, the complexity of passengers' destination was considered, and the Gini coefficient was used to quantify the inequality of passengers' destinations. With a larger Gini coefficient, the passengers' destinations were more concentrated, and thus when emergency occurs, it is more conveniently dealt with. Furthermore, from their research on major passenger destinations, and considering the Gini coefficient of passenger flow distribution, they suggested using corresponding strategies and measures by analyzing impacts and Gini coefficient when dealing with sudden increases in passenger flow.

Major passenger destinations and major driver destinations are both important in traffic analysis. In addition, the concept of "source" is used not only in transportation but also in various complex networks in terms of supply and demand, such as the Internet and information networks.

"Driver source" has important significance in urban traffic planning and traffic management, and we believe that methods utilizing it can be applied to a wider range of transport networks in the future. In the present paper, we give a summary of existing data, methods, and applications of traffic source prediction. In the current information era, abundant and highly accurate data provide the basis for travel demand forecasting. Different traffic assignment methods can be selected according to the different traffic networks. From understanding the usage of road networks, we can determine the major "driver sources" and "congested driver sources" and, moreover, can apply appropriate ways of solving traffic problems based on them. With further advances in this research, we believe that the concept of the "driver source" will be more and more widely used in transportation analysis and become a topic of greater interest.

References

1. P. Wang, T. Hunter, A.M. Bayen, K. Schechtner, M.C. González, Understanding road usage patterns in urban areas. Sci. Rep. **2**(12), 1001 (2012)
2. J. Wang, D. Wei, K. He, H. Gong, P. Wang, Encapsulating urban traffic rhythms into road networks. Sci. Rep. **4**(7488), 4141 (2014)
3. J. Wang, Y. Li, J. Liu, K. He, P. Wang, Vulnerability analysis and passenger source prediction in urban rail transit networks. PLoS One **8**(11), e80178 (2013)
4. K. He, Z. Xu, P. Wang, L. Deng, Congestion avoidance routing based on large-scale social signals. IEEE Trans. Intell. Transp. Syst. **17**, 1–14 (2016)
5. J. Wang, Q. Tan, P. Wang, Analysis of surging passenger flow in urban rail transit network. J. Railw. Sci. Eng. **12**(1), 196–202 (2015)
6. K. He, Z. Xu, P. Wang, A hybrid routing model for mitigating congestion in networks. Phys. A Stat. Mech. Appl. **431**, 1–17 (2015)
7. K. He, J. Wang, L. Deng, P. Wang, in Congestion Avoidance Routing in Urban Rail Transit Networks. *IEEE, International Conference on Intelligent Transportation Systems* (IEEE, 2015), pp. 200–205
8. M.C. González, C.A. Hidalgo, A.L. Barabási, Understanding individual human mobility patterns. Nature **453**(7196), 779–782 (2008)
9. C. Song, Z. Qu, N. Blumm, A.L. Barabási, Limits of predictability in human mobility. Science **327**(5968), 1018 (2010)
10. P. Deville, C. Linard, S. Martin, M. Gilbert, F.R. Stevens, A.E. Gaughan, et al., Dynamic population mapping using mobile phone data. Proc. Natl. Acad. Sci. U. S. A. **111**(45), 15888 (2014)

11. K.A.S.A. Khateeb, J.A.Y. Johari, W.F.A. Khateeb, Dynamic traffic light sequence algorithm using rfid. J. Comput. Sci. **4**(7), 517–524 (2008)
12. L. Yang, Detection of real-time event in intelligent traffic system based on rfid. Adv. Mater. Res. **926-930**, 1314–1317 (2014)
13. W. Wen, An intelligent traffic management expert system with RFID technology. Expert Syst. Appl. **37**(4), 3024–3035 (2010)
14. Y. Wei, M.C. Chen, Forecasting the short-term metro passenger flow with empirical mode decomposition and neural networks. Transp. Res. Pt. C Emerg. Technol. **21**(1), 148–162 (2012)
15. L. Sun, D.H. Lee, A. Erath, X. Huang, in Using Smart Card Data to Extract Passenger's Spatio-Temporal Density and Train's Trajectory of MRT System. *ACM SIGKDD International Workshop on Urban Computing ACM SIGKDD International Workshop on Urban Computing*, vol 4 (ACM, 2012), pp.142–148)
16. M.S. Iqbal, C.F. Choudhury, P. Wang, M.C. González, Development of origin–destination matrices using mobile phone call data. Transp. Res. Pt. C Emerg. Technol. **40**(1), 63–74 (2014)
17. H. Gong, L. Sun, P. Wang, Using mobile phone data to estimate trip distribution of urban mega traffic analysis zones: a case study in San Francisco. Urban Transp. China **14**(1), 37–42 (2016)
18. E.W. Dijkstra, A note on two problems in connexion with graphs. Numer. Math. **1**(1), 269–271 (1959)
19. M. Chen, A.S. Alfa, A network design algorithm using a stochastic incremental traffic assignment approach. Transp. Sci. **25**(3), 215–224 (1991)
20. M. Frank, P. Wolfe, An algorithm for quadratic programming. Nav. Res. Logist. **3**(1-2), 95–110 (1956)
21. S. Çolak, A. Lima, M.C. González, Understanding congested travel in urban areas. Nat. Commun. **7**, 10793 (2016)
22. G. Li, S.D.S. Reis, A.A. Moreira, S. Havlin, H.E. Stanley, J.S. Andrade Jr., Towards design principles for optimal transport networks. Phys. Rev. Lett. **104**(1), 018701 (2010)
23. P. Crucitti, V. Latora, S. Porta, Centrality measures in spatial networks of urban streets. Phys. Rev. E **73**(3), 036125 (2006)
24. Y. Tong, L. Lu, Z. Zhang, Y. He, W. Lu, Vulnerability of long-distance bridges and tunnels in urban roadway networks. J. Eng. Sci. Technol. Rev. **10**(1), 123–130 (2017)
25. B. Yang, W. Guo, B. Chen, G. Yang, J. Zhang, Estimating mobile traffic demand using Twitter. IEEE Wireless Commun. Lett. **5**(4), 380–383 (2016)
26. G. Karoń, Travel Demand and Transportation Supply Modelling for Agglomeration without Transportation Model, in *International Conference on Transport Systems Telematics*, (Springer, Berlin, 2013), pp. 284–293
27. X. Hu, Y.C. Chiu, J. Shelton, Development of a behaviorally induced system optimal travel demand management system. J. Intell. Transp. Syst. **21**(1), 12–25 (2017)
28. J. Wang, Q. Gu, J. Wu, G. Liu, Z. Xiong, in Traffic Speed Prediction and Congestion Source Exploration: *A Deep Learning Method. Data Mining (ICDM), 2016 IEEE 16th International Conference on* (IEEE, 2016), pp. 499–508
29. H. Gong, C. Wang, Y. Qu, K. He, P. Wang, in Locating Traffic Information Sources in Urban Areas. *Cota International Conference of Transportation Professionals.* (2016), pp. 141–151
30. J.L. Toole, S. Colak, B. Sturt, L.P. Alexander, A. Evsukoff, M.C. González, The path most traveled: travel demand estimation using big data resources. Transp. Res. Pt. C Emerg.Technol. **58**, 162–177 (2015)

Chapter 6
Analyzing the Spatial and Temporal Characteristics of Subway Passenger Flow Based on Smart Card Data

Xiaolei Ma, Jiyu Zhang, and Chuan Ding

6.1 Introduction

Passenger flow is a core feature of rail transit stations and is strongly influenced by the land utilization around the station. Therefore, the rail transit system can be improved by understanding the passenger characteristic flow and analyzing the correlation between peak passenger flow and land-use density.

In recent years, many studies have attempted to elucidate the passenger flow characteristics of rail transit systems. Sun et al. [1] applied a Bayesian algorithm to investigate the passenger distribution of complex subway networks. Zhou and Han [2] analyzed the influence of train capacity restriction on subway passenger flow under a timetable. Recognizing the rapid development of information technology, Wang et al. [3], Ma et al. [4], and Liu et al. [5] utilized big data to excavate the law of passenger flow. For example, the mobile phone data of passengers were used to summarize passenger flow and evaluate the subway service level [6].

Land use is one of the factors contributing to passenger flow. The relationship between land-use types and traffic demands has been extensively explored in the literature. Cellular automation and Markov models were used to understand the changes at regional scale, and discrete choice models to predict the changes at local level [7]. Dill [8] discovered that the rail transit convergence on the choice of subway passengers is greater than the subway location. Several scholars have proposed various evaluation systems for land use and subway passenger flow to analyze the link between the two factors [9, 10].

X. Ma (✉) · J. Zhang · C. Ding
School of Transportation Science and Engineering, Beijing Key Laboratory for Cooperative Vehicle Infrastructure System and Safety Control, Beihang University, Beijing, China
e-mail: xiaolei@buaa.edu.cn; cding@buaa.edu.cn

© Springer International Publishing AG, part of Springer Nature 2019   121
S. V. Ukkusuri, C. Yang (eds.), *Transportation Analytics in the Era of Big Data*,
Complex Networks and Dynamic Systems 4,
https://doi.org/10.1007/978-3-319-75862-6_6

Although researchers have gained knowledge on the relationship between peak passenger flow and land-use density, the following critical issues should still be addressed:

1. The characteristics of passenger flow are mostly analyzed from a macro perspective, ignoring the micro point of view. In addition, most studies considered station-level passenger flow as a static data source, thereby disregarding the temporal characteristic of passenger flow.
2. The majority of studies on the relationship between passenger flow and land-use density are qualitative, indicating the lack of quantitative calculations.

Therefore, the contributions of the present study can be summarized as follows:

1. Subway passenger flow is analyzed at station level. Considering the large number of subway stations and the temporal characteristic of passenger flow, we develop a sequential K-means clustering algorithm that utilizes smart card data to categorize subway stations in Beijing.
2. The relation between peak-hour passenger flow and land-use density is quantitatively calculated. In view of the spatial nonstationarity of the passenger flow of rail transit stations, this study proposes the geographically weighted regression (GWR) model to compute the correlation between peak passenger flow and land-use density.

The rest of the paper is organized as follows. In Sect. 6.2, a sequential K-means clustering algorithm is developed for categorizing subway stations in Beijing to analyze the differences among the station categories in terms of station-level passenger flow. In Sect. 6.3, we determine the correlation effect between peak-hour passenger flow and land-use density. We also analyze the spatial distribution of the correlation coefficients. Finally, the drawn conclusions and recommended future study directions are elaborated in Sect. 6.4.

6.2 Subway Station Classification

6.2.1 Sequential K-Means Clustering Algorithm

The development of urban rail transits in recent years has resulted in severely overcrowded passenger flow. To address this issue, the characteristics of passenger flow must be analyzed. At present, most public transit agencies have adopted smart cards, which can be used by transit planners and researchers to recognize the inherent law of passenger flow. For this purpose, subway stations are clustered because of the large number of stations and the similarity among them.

Each smart card transaction record includes an individual's boarding timestamp, alighting timestamp, and station information. The station-level smart card transaction data are typical time series data. Thus, the temporal characteristics of daily inbound and outbound subway passenger flows must be considered in the clustering. Sequential K-means clustering algorithm is an extension of traditional K-means algorithm and is specifically designed to cope with time series data [11, 22].

Time series clustering algorithms can be divided into two categories. The first category is for modifying traditional algorithms into cluster time series data. The other category is for adjusting the structure of the input data and making the modified data structures suitable for the existing clustering algorithms. According to the classification of time series clustering algorithms, the time series clustering methods include Raw-data-based and feature-based approaches [12].

Conventional clustering algorithms with suitable distance functions are adopted to the Raw-data-based method based on raw time series data [13]. One key component in model-based methods is to measure the similarity between two time series. The most commonly used distance is Euclidean distance. In addition to Euclidean distance, distance related Pearson's correlation coefficient [14], short time series (STS) distance [15], dynamic time warping (DTW) distance, and probability-based distance [16] are also applied to calculate the similarity. Košmelj and Batagelj [17] modified the relocation clustering procedure for time series data [17]. Liao et al. [18] applied K-means, fuzzy c-means, and genetic clustering methods to process time series data. Shumway [19] developed a time series clustering method based on Kullback–Leibler discrimination information measures.

For the feature-based approaches, feature vectors are extracted from the raw time series data for reducing the dimensionality of complex data sets. Wang et al. [20] selected average, standard deviation, skewness, kurtosis, and other characteristics to cluster time series data by hierarchical clustering. Alonso et al. [21] also applied the hierarchical clustering method to analyze the carbon dioxide emissions. The selected distance was probability-based distance function.

How to choose a proper approach is crucial for time series clustering [11, 22]. Unfortunately, there is no unified standard to judge the clustering result quality. Therefore, the selection of a suitable method will rely on the practical problem and data characteristics. If the data size is moderate and abundant information can be used to determine the model parameter, the Raw-data-based methods can be applied due to the ease for implementation. However, if the raw data is high-dimensional and complex, the feature-based approach is more appropriate since the number of dimensions can be significantly reduced with feature extraction, leading to an improvement of computational efficiency. In this study, the raw data of this study is ridership of Beijing metro station. The data structure is not complicated and can be well handled by traditional clustering methods. In addition, the dimension of ridership time series data at each metro station is equal. Therefore, it is suitable to apply the conventional distance to measure the similarity between different stations. Based on the above reasons, the sequential K-means algorithm is adopted in this study.

The sequential K-means algorithm belongs to the second category, as this algorithm converts time series data into longitudinal data that can be processed by traditional K-means algorithms. R provides a package for managing and calculating longitudinal data.

6.2.1.1 Algorithm Introduction

K-means is a hill-climbing algorithm under the EM class. EM algorithms work as follows. Each observation is initially assigned to a cluster. The optimal clustering is then reached by alternating two phases. During the expectation phase, the centers of the different clusters (known as seeds) are computed. During the maximization phase, each observation is then assigned to its nearest cluster. The two phases are alternated repeatedly until the clusters no longer exhibit any changes [23].

Let S be a matrix, each row represent an object, and each column represent measurements over time. $y_i = (y_{i1}, y_{i2}, \ldots, y_{it})$ is called a trajectory, and y_{ij} is the value of trajectory i measured at time j.

$$S = \begin{pmatrix} y_{11} & y_{12} & \cdots & y_{1t} \\ y_{21} & y_{22} & \cdots & y_{2t} \\ \vdots & \vdots & \vdots & \vdots \\ y_{n1} & y_{n2} & \cdots & y_{nt} \end{pmatrix} \tag{6.1}$$

The aim of clustering is to divide S into K homogeneous subgroups. The main issue in homogeneous subgroups is calculating the distance among individuals. K-means algorithm can operate with different kinds of distances, such as Manhattan, Minkowski, and Euclidean. We define $d(y_m, y_n)$ as the distance between y_m and y_n. Let Dist be a distance function and $\| \cdot \|$ be a norm. To compute the distance d between y_m and y_n, for each fixed j, we define the distance between y_{mj} and y_{nj} (distance between the individuals' state at time j) as $d_j.(y_{mj}, y_{nj}) = \text{Dist}(y_{mj}, y_{nj})$. This value is the distance between column j in matrix y_m and column j in matrix y_n. The result is a "vector of t distances" as follows:

$$(d_1. (y_{m1}, y_{n1}), d_2. (y_{m2}, y_{n2}), \ldots, d_t. (y_{t2}, y_{t2})). \tag{6.2}$$

Then, we combine these t distances by use of a function that algebraically corresponds to a norm $\| \cdot \|$ of the vector of the distance. Finally, the distance between y_m and y_n is expressed as

$$d(y_m, y_n) = \|\text{Dist}(y_{m1}, y_{n1}), \text{Dist}(y_{m2}, y_{n2}), \ldots, \text{Dist}(y_{mt}, y_{nt})\|. \tag{6.3}$$

The choice of the norm $\| \cdot \|$ and the distance Dist can lead to the derivation of a large number of distances. We can define the distance between two joint variable trajectories as

$$\text{Dist}(y_m, y_n) = \sqrt[p]{\sum_{j}^{t} |y_{mj} - y_{nj}|^p}. \tag{6.4}$$

The Euclidean distance is obtained by setting $p = 2$. The Manhattan distance is acquired by setting $p = 1$. The maximum distance is determined by passing to the limit $p \to +\infty$.

K-means computes the objective function after each iteration. S_i refers to the set of individuals belonging to category i, and c_i denotes the mean of S_i.

$$J(Y, C) = \sum_{i=1}^{k} \sum_{y_j \in S_i} d\left(y_j, c_i\right) \tag{6.5}$$

6.2.1.2 Choosing the Optimal Number of Clusters

An unresolved problem of K-means is determining the optimal number of clusters. A possible solution is to run the K-means algorithm with varying initial number of seeds and then select the "best" number of clusters according to a few quality criteria. A "good" partition indicates that the individuals within the same category are adjunct to one another as close as possible, whereas individuals under different categories are apart from each other as far as possible. The quality criteria require high values for partitions of high quality, and low values otherwise (or vice versa, depending on the criteria). Thus, most indices calculate the "within-cluster compactness index" and "between-cluster spacing index," wherein one of which is divided by the other.

Let n_m be the number of individuals in cluster m, S_i be the mean trajectories of cluster m, and $\overline{y}$ be the mean trajectories of whole individuals. The between-cluster covariance matrix is

$$B = \sum_{m=1}^{k} n_m \left(\overline{y_m} - \overline{y}\right) \left(\overline{y_m} - \overline{y}\right)'. \tag{6.6}$$

Trace(B) designates the sum of the diagonal coefficients of B. High values of *trace(B)* denote well-separated clusters, whereas low values of *trace(B)* indicate that the clusters are close to one another. The within-cluster covariance matrix W is similarly defined as follows:

$$W = \sum_{m=1}^{k} \sum_{k=1}^{n_m} n_m \left(y_{mk} - \overline{y_m}\right) \left(y_{mk} - \overline{y_m}\right)'. \tag{6.7}$$

Low values of *trace(W)* correspond to compact clusters, whereas high values of *trace(W)* correspond to heterogeneous groups.

The quality criterion combines the within and between matrices to evaluate the quality of the partition. Specifically,

Calinski and Harabasz criterion

$$C(k) = \frac{trace(B)}{trace(W)} \cdot \frac{n-k}{k-1} \tag{6.8}$$

Calinski and Harabasz, Kryszczuk variant

$$C_K(k) = \frac{trace(B)}{trace(W)} \cdot \frac{n-1}{n-k} \qquad (6.9)$$

Calinski and Harabasz, Genolini variant

$$C_G(k) = \frac{B}{W} \cdot \frac{n-k}{\sqrt{k-1}} \qquad (6.10)$$

The correct solution cannot always be obtained. In practice, researchers often resort to several criteria to strengthen the reliability of the result. This study uses Calinski and Harabasz criterion as the main criterion, while the other criteria are available for checking the result consistency.

6.2.1.3 Initialization of K-Means

The first step of the K-means algorithm is to determine the initial k cluster centers. (1) Initialization influences the convergence (local or global maximum) and (2) the computational efficiency of the algorithm. If an initialization method can determine k individuals that are fairly close to the best partition, then K-means can rapidly converge to the optimal solution. Several authors have proposed various initialization methods. The objective of initialization is to choose initial centers that are as distant as possible from one another. Such individuals ideally belong to different clusters, thereby enhancing the convergence speed. Unfortunately, no initialization method can guarantee that the optimal initial centers can be found. Thus, most approaches allow users to run the method several times. Each run starts at a different initialization to ensure that one of them reaches the global maximum. The initial clustering centers can be selected using various traditional initialization methods:

1. *randomK*: k individuals are chosen randomly as initial cluster centers.
2. *randomAll*: All individuals are randomly portioned into K clusters. The mean of each cluster is the initial cluster center.
3. *maxDist*: This method is incremental. First, it selects two individuals as the first centers that are most distant from each other. Then, it adds the individual that is the farthest from the list of centers already preselected to the center group.
4. *kmeans+*: *maxDist* is more effective than random methods because the initial centers are guaranteed distant from each other. However, this method is time consuming and complex (with a complexity of $o(n^2)$) because it computes all of the distances between each two individuals. *Kmeans+* combines *maxDist* and random methods. This method follows a similar principle: the first two individuals are chosen randomly, and the other centers are added to the list of already presented centers. Thus, the matrix of distances need not be computed among all individuals. The calculation complexity is *o(nk)*, which is less than that of *maxDist* method.

6.2.2 Clustering Result

6.2.2.1 Data Processing and Parameter Calibration

The data used in this study are obtained from the Beijing AFC system, which covers 266 stations. According to the smart card transactions data, we aggregate the station-level passenger flow every 15 min. The key information includes inbound and outbound subway passenger flows, station name, time number, and line number (Table 6.1). Except for the non-service time (11:45 PM–6:00 AM), the data complexity is $o(266 \times 76)$. This study mainly focuses on station-level passenger flow. Therefore, we merge the passenger flows from several stations with the same station name but belonging to different lines. We regard the computed average passenger flows from Monday to Friday as the base data. We then convert the time series data into the format of clusterLongData on the R software [24], which can be processed by the K-means algorithm.

Determining the proper k value and initialization method is crucial to the success of the sequential K-means clustering algorithm. To consider inbound and outbound passenger flows, we perform two clustering procedures for inbound and outbound passengers and then apply cross classification to determine the final number of categories on the basis of the clustered results. Inbound and outbound are opposite flows, and their main difference is the peak-hour passenger flow. Therefore, we select the same number for inbound and outbound directions. According to the temporal characteristic of passenger flow, the subway stations are divided into the following three categories [25]:

1. Dual-peak stations: Stations that have morning and evening peaks;
2. Single-peak station: Stations that have either a morning or an evening peak;
3. No-peak station: Stations without peaks. The passenger flow fluctuates irregularly over a day.

We choose the range of 4–6 as the initial number. After several times of running using different initial numbers, the optimal number is determined by contrasting the Calinski and Harabasz criterion. We use Euclidean distance to calculate the distance among individuals. The complexity of the cluster data is 266×76, and the runtime

Table 6.1 Example of basic data

ID	30,315	30,330	30,379
Data	20,150,602	20,150,602	20,150,602
Line number	4	4	13
Station number	29	47	43
Time	70	20	52
Inbound passenger flow	385	0	74
Outbound passenger flow	147	2	105
Line name	4 line	4 line	13 line

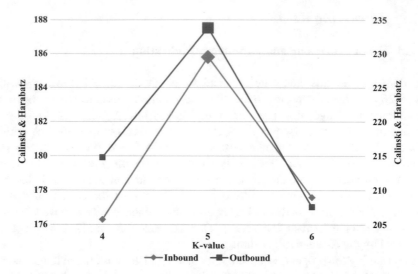

Fig. 6.1 The result of Calinski and Harabasz criterion

is within a reasonable range. Therefore, we choose *maxDist* as the initialization method to guarantee that the initial centers are as distant as possible from each other.

6.2.2.2   Weekday Station Classification Results

The algorithm is run 20 times for 4–6, and the average Calinski and Harabasz criterion is calculated (Fig. 6.1). The max Calinski and Harabasz criterion is $k = 5$ in inbound and outbound clusters. Thus, the inbound and outbound passenger flows are classified into five clusters. The inbound classifications are marked *A–E*, and the outbound classifications are marked *a–e*. Finally, the subway stations are divided into 10 groups (Table 6.2) by combining the inbound and outbound cluster results, such as *Aa* (inbound passenger belonging to cluster *A* and outbound passenger belonging to cluster *a*). The number of groups is less than 25 (5 × 5), indicating that the inbound and outbound passengers are related to each other.

The stations are divided into 10 groups marked *Xy*, where *X* and *y* represent the inbound and outbound cluster results, respectively (Fig. 6.2). The features of the different stations are summarized below.

1. *Group Aa*

   The *Aa* group is the largest category with 116 stations, accounting for 43.6%. The stations are typical dual-peak stations with morning and evening peaks in inbound and outbound directions. The morning peak passenger flow volume is around 300 people/15 min, and the evening peak passenger flow volume is around 210 people/15 min. The stations within this group are located mostly in

Table 6.2 Information of station groups

Group no.	Number of stations	Representative station name
1	116	Anhuaqiao (AHQ) Anlilu (ALL) Baliqiao (BLQ)
2	16	Andingmen (ADM) Anzhenmen (AZM) Beijing zoo
3	15	Anheqiaobei (AHQB) Beigongmen (BGM) Beiyuan (BY)
4	15	Changchunjie (CCJ) Guloudajie (GLDJ) Jintailu (JTL)
5	33	Babaoshan (BBS) Bajiao (BJ) Gongyixiqiao (GYXQ)
6	32	BaiShiQiaoDong (BSQD) Beijing South Railway Station
7	22	Caoqiao (CQ) Dongdaqiao (DDQ) Fuxingmen (FXM)
8	9	Caofang (CF) Huilongguan (HLG) Huoying (HY)
9	5	Dawanlu (DWL) Dongzhimen (DZM) Xizhimen (XZM)
10	3	Chaoyangmen (CYM) Guomao (GM) Xierqi (XEQ)

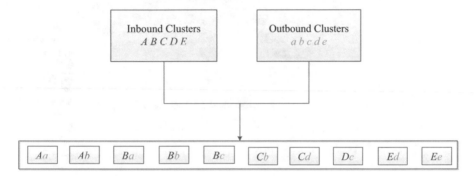

Fig. 6.2 Cross classification

regions of the Beijing rail transit network, such as stations in suburban areas, where passengers primarily commute between their residence and workplace.

The passenger flow characteristics of the *Aa* group are below the designed station capacity. Nearly half of the Beijing subway falls within this category. These stations are unlikely to cause congestion from the low passenger flow volume. Thus, the main problem of Beijing subway stations is not the oversaturated passenger flow but the reasonable passenger flow allocation to those stations that are underutilized (Figs. 6.3 and 6.4).

2. *Group Ab*

The *Ab* group contains 16 stations, accounting for 6%. The *Ab* group belongs to employment-oriented stations. The inbound evening peak passenger flow volume is 720 people/15 min, and the outbound morning peak passenger flow volume is 1050 people/15 min. The passenger flow manifests as "morning–outbound and evening–inbound," which is the core feature of employment-oriented stations. The max inbound and outbound metro ridership are 16,200 and 18,000 people/day, respectively. Compared with the *Aa* group, the *Ab group* has a considerably larger average passenger flow, and the increase in peak passenger flow is much higher than that in the *Aa* group. However, the increase in passenger flow does not cause obvious queuing (Fig. 6.5).

Fig. 6.3 Spatial distribution
of subway station categories

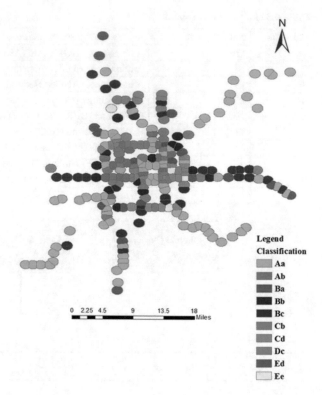

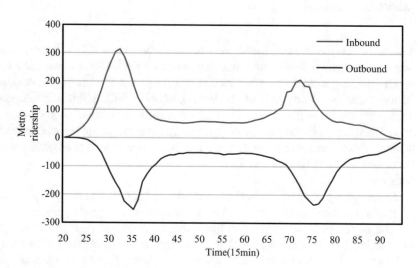

Fig. 6.4 Temporal distribution of passenger flow (*Aa* group)

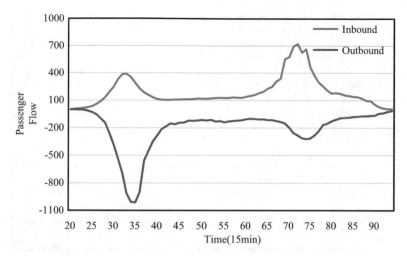

Fig. 6.5 Temporal distribution of passenger flow (*Ab* group). The *Ab* group stations are mainly located downtown, primarily in the north. The main land types surrounding the stations are employment land with a few residential buildings. The intensity of land development surrounding the *Ab* group stations is higher than that surrounding the *Aa* group

3. *Group Ba*

The *Ba* group contains 15 stations, accounting for 5.6%. The temporal characteristics of the inbound and outbound passenger flows are contrary to that in the *Ab* group. The passenger flow manifests as "morning–inbound and evening–outbound," which is the direction of residence-oriented stations. The inbound morning peak passenger flow volume is 1000 people/15 min, which is larger than the outbound evening peak. Commuters enter stations to go to their workplaces during morning peak hours and leave stations to go to their residences during evening peak hours. The max inbound and outbound metro ridership are 17,800 and 15,800 people/day, respectively. The level of the *Ba* group passenger flow volume is within the same range as the *Ab* group, suggesting that "employment" and" residence" are a pair of opposite directions for commuters. The *Ba* group stations are distributed in the suburbs outside of the 4th Ring Road in Beijing. These regions are surrounded by a high density of residential buildings (Fig. 6.6).

4. *Group Bb*

The *Bb* group contains 15 stations, accounting for 5.6%. The characteristics of this group are similar to that of the *Aa* group. In particular, double peaks of inbound and outbound passenger flows are magnified in equal proportions in this group. The inbound and outbound morning peak metro ridership reach 1160 people/15 min. The max inbound and outbound metro ridership are 47,000 and 41,000 people/day, respectively.

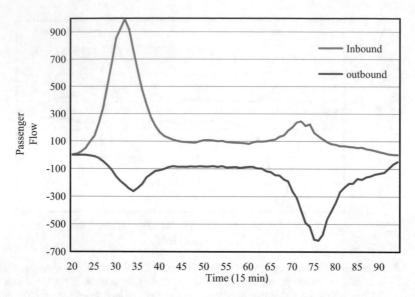

Fig. 6.6 Temporal distribution of passenger flow (*Ba* group)

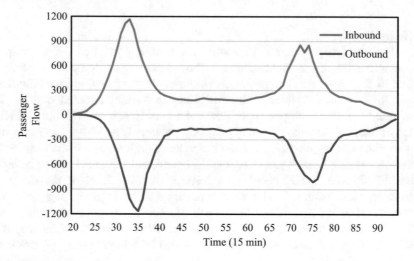

Fig. 6.7 Temporal distribution of passenger flow (*Bb* group)

The *Bb* group stations are dispersedly distributed without clear spatial distribution regularities and are located closer to the downtown than the *Aa* group (Fig. 6.7).

5. *Group Bc*

The passenger flow in the *Bc* group is magnified in equal proportion as that in the *Ba* group. This group is a representative residential station wherein residential-direction peak passenger flows are three times the employment-

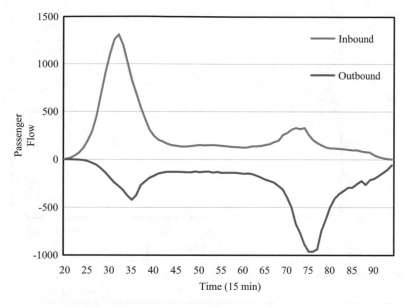

Fig. 6.8 Temporal distribution of passenger flow (*Bc* group)

direction peak passenger flows. The inbound morning and outbound evening peak metro ridership are 1300 and 1000 people/15 min. These stations are located downtown and are closer to the city center than the *Ba* group. The land-use areas around these stations are residence communities, which attract a large number of passengers. We find that most stations in the *Bc* group will undergo congestion when the peak passenger flow volume reaches the level of the *Bc* group (Fig. 6.8).

6. *Group Cb*

The *Cb* group contains 32 stations, accounting for 12%. Although the passenger flow is employment-directed, the differences between the residential-direction and the employment-direction are not as striking as those in the *Ab* group. The inbound evening and outbound morning peak metro ridership are 1200 and 1540 people/15 min. The max inbound and outbound metro ridership are 40,000 and 42,000 people/day (Fig. 6.9).

Most stations are located in the region of 4th Ring Road, and they are more concentrated and closer to the downtown than the *Ab* group. External hub stations, such as the Beijing South, Beijing, and Beijing West Railway Stations, are assigned to the *Cb* group. External hub subway stations have no significant peak passenger flows.

The characteristics of the other groups, such as *Cd* and *Dc*, are similar to those of the abovementioned six groups. Comparison reveals that the different groups possess resemblances among one another. The 10 groups are classified under three categories: employment-oriented, dual-peak, and residence-oriented stations.

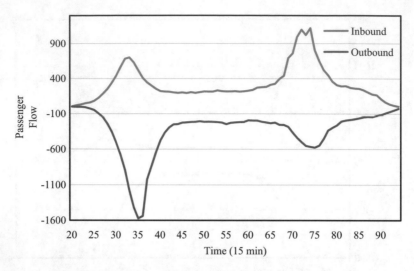

Fig. 6.9 Temporal distribution of passenger flow (*Cb* group)

Dual-peak stations are stations with double peaks in the inbound and outbound passenger flows. The gross passenger flow volume is at the middle and lower levels, avoiding any crowding phenomenon. Many stations have excess capacity that causes wastage of space. The land-use types surrounding the stations are manifold, and these stations are widely distributed without centers.

The land-use type surrounding employment-oriented stations mainly comprises employment land. Passenger flows have obvious directivity and are divided into peak and off-peak directions. Inbound evening peak and outbound morning peak belong to employment peak direction, whereas the others are assigned to employment off-peak direction. Under this commuting direction, passengers around employment-oriented stations leave the stations to go to their workplaces in the morning and enter the stations during the evening peak hours. These stations are located around employment centers, such as CBD and Financial Street.

The peak and off-peak directions of residence-oriented stations are opposite to that of employment-oriented stations. The inbound evening peak and outbound morning peak belong to residence off-peak direction. Commuters enter stations in the morning and leave the stations to return to their residences in the evening. Residence-oriented stations are located in residential centers away from the downtown, such as HLG. Most passengers live in this area because of the low prices. The difference between peak and off-peak direction passenger flows is more significant in residence-oriented stations than that in employment-oriented stations.

Sequential K-means algorithm is used to divide subway stations in Beijing into 10 groups under three categories. We analyze the differences of these categories in terms of station-level passenger flow. We also examine the relation between passenger flow and land use to verify the clustering results.

6.2.2.3   Weekend Passenger Flow Clustering

The aforementioned clustering results are based mainly on weekday passenger data. Thus, we contrast weekend passenger flow clustering with that based on weekday passenger flow. With the use of the same algorithm, inbound and outbound passenger flows are both divided into two categories. Similar to the weekday data, weekend inbound clustering results are marked *A*, *B*, and outbound clustering results are marked *a*, *b*. The main distinction of the two categories is that the passenger flow volume of *Aa* stations is larger than that of *Bb* stations. If the inbound cluster results are combined with the outbound cluster results, the stations can be divided into four groups (*Aa, Ab, Ba, Bb*) according to the weekday passenger data. The distributions of weekend passenger flow are stable and have no obvious peak flows. The characteristic of weekend passenger flow proves that most Beijing subway riders are commuters.

The *Bb* groups include 32 stations, accounting for 12%. The *Bb* group stations, such as XD and WFJ, among others, are major hubs of the Beijing subway. These stations belong to the groups with large passenger flows in the weekdays' clustering results. The max passenger flow volume ranges from 400 to 500 people/15 min, which does not lead to crowding (Fig. 6.10).

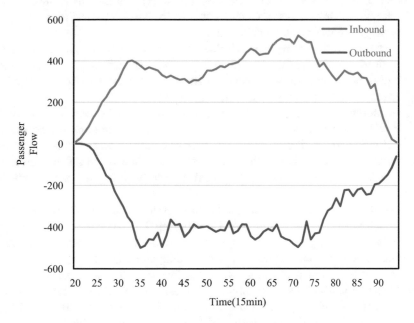

Fig. 6.10 Temporal distribution of weekend passenger flow (*Bb* group)

6.3 Correlation Between Passenger Flow and Land Use

In the previous chapter, we develop a sequential K-means clustering algorithm for categorizing Beijing subway stations. The fluctuation of station-level passenger flow is strongly influenced by its surrounding land-use types. Thus, in this section, we use GIS data to determine the correlation effect between peak-hour passenger flow and land-use density.

6.3.1 Calculation of Land-Use Density

To compute land-use density, we should confirm the influential region of a station. Radiation range is the station-centered region with respect to transportation connectivity. Buffer and Thiessen polygon are the two most commonly employed methods for generating the influential region.

Buffer refers to the services range of geographically spatial objects. It builds a fixed-width region surrounding a point, a line, and a plane. Buffer is an important method in geographical information system and is widely used in spatial analysis. Thiessen polygons consist of the perpendicular bisector of the line connecting two adjacent points. This method guarantees that every point is closest to central point of the polygon to which the point belongs.

Buffer and Thiessen polygon are limited by certain drawbacks. Buffer influential regions overlap with other regions belonging to multiple buffers. Thiessen polygons do not pose range limitations, especially when the points are located around edges. Such setting is inconsistent with practical situations.

To overcome the limitations of these methods, we calculate the influential region by use of the two methods separately and then consider the common portions as the final influential region. In this way, different stations do not overlap, and the distance between two points in the same buffer does not exceed a fixed value. The buffer radius is 1000 m, which is a typical average walking distance for subway riders (Fig. 6.11).

The GIS map contains all land-use types in Beijing. For simplicity, we choose key variables affecting passenger flow. Considering the practical circumstances, we extract 11 land-use types: restaurants, companies, bus stops, parking lots, financial institutions, research and education, retail, commercial buildings, entertainment, medical services, and residential buildings. We map the land-use data on the influential region to compute the number of land-use types surrounding the stations. The quantity is divided by the influential region area to obtain the density of every land-use type. The details of land-use variables are showed in Table 6.3.

Station-level passenger flow is a time series data that cannot be a dependent variable of regression analysis. The most important parameter of station-level passenger flow is the metro ridership during peak hours, which are static data that can be a dependent variable. Thus, we evaluate the correlation effect between peak-hour passenger flow and land-use density.

Fig. 6.11 Influential region

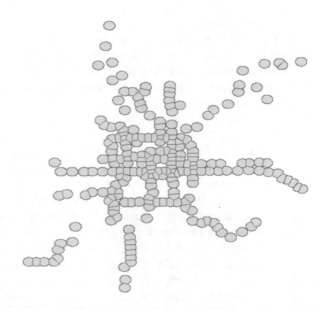

Table 6.3 Land-use variables

Type	Variable	Description
Land use	Residential building density	Number of residential buildings per km^2
	Place of employment density	Number of companies, agencies, and government agencies per km^2
	Commercial building density	Number of shopping malls per km^2
	Restaurant density	Number of restaurants per km^2
	Entertainment density	Number of entertainment centers per km^2 in each TAZ
	Financial institution density	Number of financial institutions per km^2
	Retail density	Number of retail establishments per km^2
	Research and education density	Number of universities and educational services per km^2
	Medical service density	Number of medical institutions per km^2
Transport	Bus stop density	Number of bus stops per km^2
	Parking lot density	Number of parking lots per km^2

6.3.2   Differences in Land-Use Density Between Different Group Stations

We divide Beijing subway stations into three categories. The fluctuation of station-level passenger flow is strongly influenced by its surrounding land-use types, as land use is the primary cause of trip generation and attraction. Thus, we compare the land-use density between different groups.

1. *Aa* group and *Bb* group

    The *Aa* and *Bb* groups are dual-peak stations demonstrating similar station-level fluctuations. Their main distinction is passenger flow volume. The land-use density surrounding the *Bb* group is larger than that surrounding the *Aa* group. The increase in passenger flow is attributed to the increase in all types of land-use density rather only a single type.

2. *Cd* group and *Dc* group

    The *Cd* and *Dc* are employment-oriented stations exhibiting similar passenger flow characteristics. The stations in the *Dc* group display a greater commuting passenger flow than those in the *Cd* group. Both groups have similar land-use densities. Compared with the other groups, the companies in the *Dc* and *Cd* groups have greater land-use densities. The company land-use density in *Cd* is 80 per square kilometer, while that in *Dc* is 140 per square kilometer. The residential land-use density remains unchanged. The residence passenger flow in the *Dc* group is larger than that in the *Cd* group, and the difference between company land-use density and residential land-use density is more significant between the two groups. This finding implies that the residence feature of a station is related to the difference between company land density and residential density.

The station's land-use density in the same category has the same distribution. Two key indicators are employment density and the difference between employment and residential densities. The former reflects the employment characteristics, and the latter indicates the residential characteristics of a station. A single residential density is inadequate for identifying station-level residential characteristics because detailed information on each residential community, such as house price and floor area ratio, are lacking.

6.3.3 Geographically Weighted Regression Model

The different groups differ most significantly in peak-hour passenger flow, which is strongly influenced by the land-use density around stations. Peak-hour passengers are measured depending on the spatial position of the stations. The variable displays different structures and characteristics with changes in the locations. The phenomenon that the relationships and structures of the variable change with geographic positions is called spatial nonstationarity. For example, we compute the regression functions between peak-hour passenger flow y and land-use density x_1, x_2, ..., x_p. Basic data change with the geographic position of the observation points. These changes are highly complex and cannot be described by a specific function.

In the metro system, ridership estimation models are crucial for the analysis of project feasibility [26]. In the past decades, the ridership models have been dominated by the four-step approach [27] but yield less effective results for forecasting ridership on more detailed scales. To address this issue, the direct models based on the multiple regression analysis are adopted to analyze the relationship between land use and ridership [28]. It is believed to better capture the effect of built environment on ridership. However, the space feature is difficult to assess prior to the investigation. If linear regression or single nonlinear regression is used to analyze the correlation, then the results are often unsatisfactory. The reason is that an important assumption of these global models is the homogeneity of the variables, which make the results reflect the "global average." Thus, we must modify the model to address this problem.

The GWR model embeds the spatial position of the data into regression parameters and uses local least square method to evaluate parameters point-by-point. The weighting factor is the distance between observations. Spatial nonstationarity can be observed from the parameter estimation of each point. Peak-hour passenger flow data are the typical spatial data with spatial nonstationarity. Thus, we select the GWR model to compute the correlation between passenger flow volume and land-use density.

6.3.3.1 Introduction to GWR

In spatial analysis, n observations are collected in n spatial positions. The GWR model can handle the spatial nonstationarity by use of different regression functions to show the spatial difference of observations. GWR is an extension of linear regression and embeds the spatial position of the data into regression parameters [29, 30]. The regression function is

$$y_i = \beta(u_i, v_i) + \sum_{k=1}^{p} \beta_k(u_i, v_i) x_{ik} + \varepsilon_i, \tag{6.11}$$

where (u_i, v_i) is the coordinate of the observation i. $\beta_k(u_i, v_i)$ is the regression parameter k of observation i. The ε_i distribution satisfies $\varepsilon_i \sim N(0, \sigma^2)$, $Cov(\varepsilon_i, \varepsilon_j) = 0 (i \neq j)$. The regression function is simplified into

$$y_i = \beta_{io} + \sum_{k=1}^{p} \beta_{ik} x_{ik} + \varepsilon_i, \quad i = 1, 2, \ldots, n. \tag{6.12}$$

If $\beta_{1k} = \beta_{2k} = \cdots = \beta_{nk}$, then GWR degrades into linear regression.
The regression equation is placed into the matrix.

$$y = (X \otimes \beta') I + \varepsilon \tag{6.13}$$

In Eq. (6.13), $\otimes$ is the logical multiplication of the matrix, wherein elements of X are multiplied by the corresponding elements of β'. n observations and p variables are used. X and β' are both $n \times (p + 1)$-dimensional matrix. I is a $(p + 1) \times 1$-dimensional unit vector. β has n groups of local regression parameters.

$$\beta = \begin{pmatrix} \beta_{10} & \cdots & \beta_{i0} & \cdots & \beta_{n0} \\ \beta_{11} & \cdots & \beta_{i1} & \cdots & \beta_{n1} \\ \vdots & \vdots & \vdots & \vdots & \vdots \\ \beta_{1p} & \cdots & \beta_{ip} & \cdots & \beta_{np} \end{pmatrix} \tag{6.14}$$

Each observation has a different set of regression parameters. Thus, the number of unknown parameters is $n \times (p + 1)$, which is greater than n. As such, unknown parameters are difficult to estimate using parametric regression.

When computing the regression coefficient of point i, the observations have varying importance. A closer distance indicates a greater importance. By using weighted least square method, the regression parameters of point i can be estimated by calculating the minimum of

$$\sum_{j=1}^{n} w_{ij} \left(y_j - \beta_{io} - \sum_{k=1}^{p} \beta_{ik} x_{ik} \right)^2, \tag{6.15}$$

where w_{ij} is the weight that is inversely proportional to the distance between points i and j. Let$\beta_t = (\beta_{io}, \beta_{i1}, \ldots, \beta_{ip})'$, $W_t = \text{diag}(w_{i1}, w_{i2}, \ldots, w_{in})$. The regression parameter estimates of point i is

$$\widehat{\beta}_i = \left(X' W_i X \right)^{-1} X' W_i y. \tag{6.16}$$

Let $C_i = (X' W_i X)^{-1} X' W_i$, then

$$Var\left(\widehat{\beta}_i \right) = C_i C_i' \sigma^2. \tag{6.17}$$

The regression value of point i is calculated as

$$\widehat{y}_i = X_i \widehat{\beta}_i = X_i \left(X' W_i X \right)^{-1} X' W_i y, \tag{6.18}$$

where X_i is the ith row vector of X. $S_i = X_i (X' W_t X)^{-1} X' W_i$ is the hat matrix of point i. $\widehat{y}_i = S_t y$. According to the hat matrix, the mathematical expectation of regression is

$$X_i E\left(\widehat{y}_i \right) = X_t \left(X' W_t X \right)^{-1} X' W_t E(y). \tag{6.19}$$

Given that

$$X' W_t E(y) = \left(X_1', X_2', \ldots, X_n' \right) \begin{bmatrix} w_{t1} & 0 & \cdots & 0 \\ 0 & w_{t2} & \cdots & 0 \\ \vdots & \vdots & \ddots & \vdots \\ 0 & 0 & \cdots & w_{in} \end{bmatrix} \begin{bmatrix} X_1 \widehat{\beta}_1 \\ X_2 \widehat{\beta}_2 \\ \vdots \\ X_n \widehat{\beta}_n \end{bmatrix} = \sum_{j=1}^{n} w_{ij} X_j' X_j \widehat{\beta}_j, \tag{6.20}$$

we obtain

$$E(\widehat{y_i}) = \sum_{j=1}^{n} w_{ij} X_i (X'W_i X)^{-1} X_j' X_j \widehat{\beta_j}. \tag{6.21}$$

Under this assumption, the error of observations is independent, i.e., $E(\varepsilon_i) = 0$, $Var(\varepsilon_i) = \sigma^2$. Thus, $Var(y) = \sigma^2 I_n$.

$$Var(\widehat{y_i}) = Var\left(X_i (X'W_i X)^{-1} X'W_i Var(y)\right)$$

$$= \sigma^2 X_i (X'W_i X)^{-1} X'W_i^2 X (X'W_i X)^{-1} X_i' \tag{6.22}$$

The residual is computed as $e_i = y_i - \widehat{y_i} = y_i - S_i y$.

The regression matrix $\widehat{\beta}$ below is determined using the above method:

$$\widehat{\beta} = \begin{pmatrix} \widehat{\beta_{10}} & \cdots & \widehat{\beta_{i0}} & \cdots & \widehat{\beta_{n0}} \\ \widehat{\beta_{11}} & \cdots & \widehat{\beta_{i1}} & \cdots & \widehat{\beta_{n1}} \\ \vdots & \vdots & \vdots & \vdots & \vdots \\ \widehat{\beta_{1p}} & \cdots & \widehat{\beta_{ip}} & \cdots & \widehat{\beta_{np}} \end{pmatrix}. \tag{6.23}$$

Each column represents the estimated value of the same regression parameter according to different observations that show the spatial nonstationarity of the variable.

The weighting function is an important feature of GWR, as the core of GWR is to consider the spatial relationship among different observations. This relationship is reflected by the weighting function. Gaussian function is the most commonly used method for computing the weight. In this study, Gaussian function is employed to fit the relationship between weight and distance. The function is

$$w_{ij} = \exp\left(-(d_{ij}/b)^2\right), \tag{6.24}$$

where b is the bandwidth that describes the function relationship between weight and distance.

6.3.3.2 Geographically Weighted Regression Implementation

According to the characteristic of station-level passenger flow, we choose four passenger flows, namely, morning peak-hour inbound passenger flow, morning peak-hour outbound passenger flow, evening peak-hour inbound passenger flow, and evening peak-hour outbound passenger flow, as dependent variables. The densities of 11 land-use types are free variables. Thus, we build four regression equations.

Table 6.4 Variable explanation

Dependent variable	Free variable
Morning peak-hour inbound passenger flow	Bus stop, parking lot, residential building
Evening peak-hour inbound passenger flow	Employment, bus stop, parking lot Commercial building
Morning peak-hour outbound passenger flow	Employment, bus stop, parking lot Commercial building
Evening peak-hour outbound passenger flow	Bus stop, parking lot, residential building, entertainment

Before calculating the regressive equation, we reduce the dimensions of the free variables. Multiple stepwise regression method can make a selection from many free variables. SPSS is utilized to select free variables for four regressive equations (Table 6.4).

Evening peak-hour inbound passenger flow and morning peak-hour outbound passenger flow have the same remaining free variables and belong to the working direction. The mutual remaining free variables of morning peak-hour inbound passenger flow and evening peak-hour outbound passenger flow are bus stop, parking lot, and residential building. In addition, the entertainment land use affects the evening peak-hour outbound passenger flow. These results indicate that peak-hour passenger flows are influenced by employment, residential building, and traffic facilities surrounding the stations.

GWR is based on the spatial nonstationarity of variables. Global Moran's I is the measure of spatial correlation, which can be expressed as

$$I = \frac{n}{S_0} \frac{\sum_{i=1}^{n} \sum_{j=1}^{n} w_{ij} z_i z_j}{\sum_{i=1}^{n} z_i^2}, \tag{6.25}$$

where z_i is the deviation between the average $(x_i - \overline{X})$ and property of element i. w_{ij} is the spatial weight between i and j. n is the number of elements. S_0 is the sum of the spatial weights.

$$S_o = \sum_{i=1}^{n} \sum_{j=1}^{n} w_{ij} \tag{6.26}$$

z_I can be calculated as

$$z_I = \frac{1 - E[I]}{\sqrt{V[I]}} \tag{6.27}$$

$E[I] = -1/(n-1)$, $V[I] = E[I^2] - E[I]^2$.

Table 6.5 Confidence interval

z_I (standard deviation)	*P-value* (probability)	Confidence coefficient (%)
<-1.65 or $>+1.65$	<0.10	90
<-1.96 or $>+1.96$	<0.05	95
<-2.58 or $>+2.58$	<0.01	99

In Eq. (6.27), z_I and *P-value* indicate the statistical significance (Table 6.5). Moran's I lies between -1 and $+1$. Moran's I > 0 indicates that the spatial relationship has a positive correlation. Correlations steadily increase with the absolute value of Moran's I. If Moran's I < 0, the spatial relationship is negatively correlated. Moran's I $= 0$ indicates the independence of the spatial data.

The confidence levels of the four dependent variables are above 99%, and all Moran's I values are greater than 0.15. These results confirm the positive correlation among the dependent variables, thereby providing the theoretical basis for further analysis of the spatial correlation.

The GWR model is sensitive to bandwidth. An unreasonable bandwidth may lead to a deviation of regression parameter estimation. In this study, the optimal bandwidth can be calculated as

$$CV = \frac{1}{n} \sum_{i=1}^{n} \left[y_i - \widehat{y_{\neq i}}(b) \right]^2, \tag{6.28}$$

where $\widehat{y_{\neq i}}$ is the other regression point except for i. ARCGIS draws the different bandwidth with the corresponding CV value into the trend line to facilitate the selection of the optimal bandwidth.

To verify the rationality of filtering free variables, we calculate the GWR model with all free variables and remaining free variables.

The R^2 values increase after some free variables are removed, demonstrating that screening free variables can simplify the model and improve the accuracy. Thus, we regard the remaining variable results as the final results. In addition, the R^2 of direct models are also showed in Table 6.6. It can be found that the R^2 of GWR are higher than direct models, indicating that the GWR accurately depicts the relationship between ridership and land use (Table 6.7).

Other parameters can be used to justify the rationality of the GWR model.

1. Condition numbers

    Condition numbers are used to evaluate local collinearity. When the condition numbers are more than 30, the fitting result may be unreliable. The condition numbers for four regressions that meet the requirements are less than 30 (Table 6.8).

2. Standard Residual (StdResid)

    High or low predicted values imply that some key explanatory variables are likely to miss in the regression model. The StdResid of GWR can assess if the model loses key variables. The significant clustering of StdResid in statistics

Table 6.6 Moran's I

Dependent variables	Morning peak-hour inbound passenger flow	Evening peak-hour inbound passenger flow	Morning peak-hour outbound passenger flow	Evening peak-hour outbound passenger flow
Moran's I	0.246622	0.366187	0.348221	0.16163
Expectation index	−0.004425	−0.004425	−0.004425	−0.004425
Variance	0.000614	0.000611	0.000614	0.000618
z_I	10.131921	14.995811	14.232787	6.681667
P-value	0	0	0	0

Table 6.7 GWR result

	Residual squares	Sigma	AIC	R^2 of direct	R^2 of GWR
Morning peak-hour inbound passenger flow (all free variable)	372,494,548.53	1563.81	4051.42	0.40	0.60
Morning peak-hour inbound passenger flow (all remaining variable)	367.748,936.27	1479.61	4006.42	0.36	0.61
Evening peak-hour inbound passenger flow (all free variable)	471,501,758.06	1567.01	4011.14	0.59	0.65
Evening peak-hour inbound passenger flow (all remaining variable)	437,886,316.38	1509.96	3992.26	0.58	0.67
Morning peak-hour outbound passenger flow (all free variable)	1,051,239,077.58	2336.11	4192.01	0.60	0.66
Morning peak-hour outbound passenger flow (all remaining variable)	897,986,063.89	2229.66	4178.87	0.59	0.71
Evening peak-hour outbound passenger flow (all free variable)	322,471,204.46	1352.46	3957.43	0.15	0.43
Evening peak-hour outbound passenger flow (all remaining variable)	308,316,004.50	1295.30	3929.39	0.13	0.45

indicates a mis-specified regression model. We compute Moran's I of StdResid, and the values are approximately equal to 0. Thus, the StdResid does not indicate any clustering center. The free variables are sufficient without the missing variables (Table 6.9).

Table 6.8 Condition number

	Max conditions	Min conditions
Morning peak-hour inbound passenger flow	12.03	4.22
Evening peak-hour inbound passenger flow	11.94	4.67
Morning peak-hour outbound passenger flow	13.71	5.24
Evening peak-hour outbound passenger flow	11.42	7.18

Table 6.9 StdResid Moran's I

	StdResid Moran's I
Morning peak-hour inbound passenger flow	−0.001177
Evening peak-hour inbound passenger flow	−0.001177
Morning peak-hour outbound passenger flow	−0.018227
Evening peak-hour outbound passenger flow	0.001277

Table 6.10 Regression coefficients of morning peak-hour inbound passenger flow

	Minimum	Upper quartile	Median	Lower quartile	Maximum	Average	Standard deviation
Bus stop	−31.03	2.76	25.78	52.46	246.33	40.13	57.55
Parking lot	−434.88	−142.21	−65.31	−31.57	44.87	−93.92	93.26
Residential building	−59.33	48.26	93.99	142.99	335.23	106.18	79.98

6.3.3.3 Regression Coefficient

1. Morning peak-hour inbound passenger flow

The characteristic values of each regression coefficient of the free variables are summed (Table 6.10).

According to the average, residential building has the greatest impact on morning peak-hour inbound passenger flow. A higher residential area density corresponds to a greater number of passengers taking the subway. A strong negative correlation exists between parking lot density and morning peak-hour inbound passenger flow. Bus stop density is positively correlated with morning peak-hour inbound passenger flow. During morning peak hours, buses and subways complement each other, whereas cars and subways compete against each other.

The correlation coefficient of residential building density varies from −31.03 to 246.33, with most coefficients being greater than 0. The variation shows that different spatial distributions significantly influence the morning peak passenger flow. Figure 6.12a depicts the spatial distribution of the residential density coefficient. A darker color indicates a larger coefficient. The outside region is darker than the downtown, particularly in the southwest and northeast areas along the Fangshan Line and the 15 Line. Passengers in these regions tend to take the subway in the morning, as it is the main transport mode in these regions. Figure 6.12b illustrates the spatial distribution of the parking lot density coefficient.

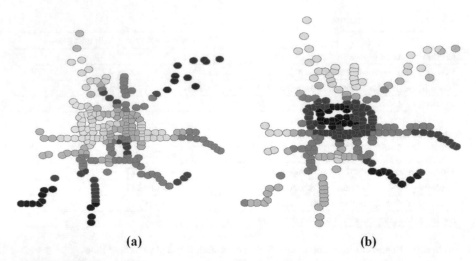

(a) (b)

Fig. 6.12 Spatial distribution of morning peak-hour inbound passenger flow: (a) residential building and (b) parking lot

Table 6.11 Regression coefficients of evening peak-hour inbound passenger flow

	Minimum	Upper quartile	Median	Lower quartile	Maximum	Average	Standard deviation
Bus stop	−97.02	−29.88	−18.91	−11.78	7.72	−23.32	18.54
Parking lot	−6.20	47.31	87.10	101.64	177.62	78.38	43.70
Commercial building	20.77	63.34	100.01	136.72	498.08	118.68	84.72
Employment	−38.51	−4.50	0.26	10.66	17.18	1.47	11.15

All coefficients are negative, such that the light color represents strong negative correlations. The correlations are distributed along the ring roads. Passengers in the suburbs prefer to drive to work. Passengers around the 1 Line and the Yizhuang Line are more likely to take the subway than drive cars.

2. Evening peak-hour inbound passenger flow

Commercial building andemployment density are positively correlated with evening peak-hour inbound passenger flow, as this passenger flow belongs to the working direction. By contrast, bus stop and evening peak-hour inbound passenger flow are negatively correlated. During the evening peak hours, buses and subways display a competitive relationship; that is, passengers choose multiple transport modes to return home or visit entertainment places. This finding conforms to the characteristics of Beijing subway passenger flow that morning peak-hour passenger flow is larger than the evening one (Table 6.11).

The bus stop density coefficients are distributed into rings (Fig. 6.13a), and the density is lighter in the downtown than in the periphery. Figure 6.13b depicts the spatial distribution of the employment density coefficient. The northeast–southwest diagonal denotes the distribution boundary. The employments in the

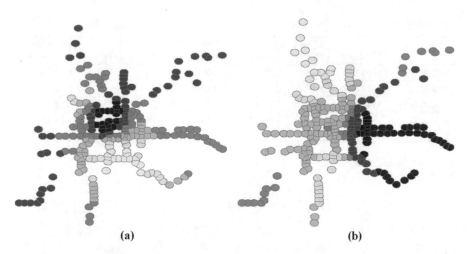

(a) (b)

Fig. 6.13 Spatial distribution of evening peak-hour inbound passenger flow: (**a**) residential building and (**b**) parking lot

Table 6.12 Regression coefficients of morning peak-hour outbound passenger flow

	Minimum	Upper quartile	Median	Lower quartile	Maximum	Average	Standard deviation
Bus stop	−217.38	−58.07	−36.07	−20.29	20.02	−46.55	42.48
Parking lot	−67.26	36.73	133.70	172.01	378.58	115.19	92.03
Commercial building	9.69	93.39	202.98	272.37	877.00	212.40	149.77
Employment	−71.19	−7.20	1.84	18.32	26.17	1.73	18.53

South–East significantly affect the evening peak-hour inbound passenger flow. A number of coefficients in the North–West are negative. This area contains many residential communities. Passengers prefer transport modes other than the subway because their workplaces are close to their residences.

3. Morning peak-hour outbound passenger flow

The average of thefour regression coefficients is similar to the evening peak-hour inbound passenger flow: positive correlation between passenger flow and parking lot, commercial, and employment; negative correlation between bus stop and passenger flow. The two passenger flows belong to the working direction, and the spatial distribution of the coefficients has similar variation patterns (Table 6.12).

4. Evening peak-hour outbound passenger flow

The coefficients of the evening peak-hour outbound passenger flow are similar to those of morning peak-hour inbound passenger flow, which both belong to living direction. Aside from bus stop, parking lot, and residential building, entertainment also affects the evening peak-hour outbound passenger flow. Entertainment is an additional free variable because passengers have sufficient time to relax after work. This difference is noticeable between the evening and morning peak-hour inbound passenger flows (Table 6.13).

Table 6.13 Regression coefficient of evening peak-hour outbound passenger flow

	Minimum	Upper quartile	Median	Lower quartile	Maximum	Average	Standard deviation
Bus stop	−10.32	2.41	20.19	30.00	117.14	25.88	32.74
Parking lot	−172.79	−82.57	−49.31	−15.39	22.92	−52.14	43.62
Entertainment	−41.00	6.63	19.15	33.00	73.95	19.80	19.85
Residential building	−42.13	12.72	60.90	87.88	143.78	53.85	44.16

Fig. 6.14 Spatial distribution of entertainment

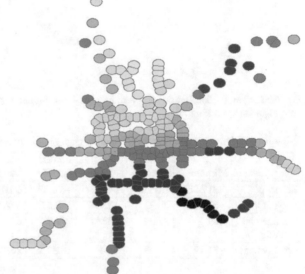

Figure 6.14 depicts the spatial distribution of entertainment density coefficients. The coefficients in the downtown are negative, suggesting that the passengers prefer other transport modes rather than the subway because the entertainment places are close to the workplace. By contrast, the coefficients are larger in the suburbs than in the downtown, indicating that passengers tend to take subway to the entertainment places around suburbs.

The correlation between peak-hour passenger flow and land-use density has obvious directionality. The regression correlations belonging to the same direction display the same patterns. Buses can either be complementary or competitive to subways depending on the time and direction of the passenger flows. The spatial distribution of the coefficients is influenced by many factors, such as urban planning scheme and subway line construction.

6.4   Conclusions

Beijing subway passenger flow is hindered by serious problems of passenger surge and unbalanced distribution, which are caused by the large population base and irrational land-use planning. Thus, this study explores the correlation between land use and passenger flow. First, we analyze the fluctuation of station-level passenger flow. Given their large number, the stations are classified. We develop a sequential K-means clustering algorithm that considers the temporal characteristics of station-level passenger flow. The stations are divided into 10 groups under three categories: employment-oriented, dual-peak, and residence-oriented stations. Peak-hour passengers of employment-oriented stations are concentrated in morning peak outbound and evening peak inbound, whereas residence-oriented stations are concentrated in the two other directions.

The clustering results indicate that most Beijing subway riders are commuters. Although Beijing has a large passenger flow, the most serious issue is the unbalanced distribution of passenger flow. This problem is closely related to land-use type. Thus, we employ the GWR model to determine the correlation effect between peak-hour passenger flow and land-use density. This model can embed the spatial position of data into the regression parameters and use local least square method to evaluate parameters point-by-point. We then analyze the spatial distribution of the correlation coefficients. According to the fitting results, peak-hour passenger flows are related to employment, residence, and transportation facilities. Morning peak-hour inbound and evening peak-hour outbound passenger flows belong to the residence direction with the same features of regression coefficient, whereas the two other types of passenger flows belong to the employment direction with the same coefficient characteristic.

The innovations of the study are as follows. (1) Sequential K-means clustering algorithm is presented to categorize the stations on the basis of smart card data; most previous studies ignored the temporal characteristics of daily inbound and outbound subway passenger flow. Passenger flow is regarded as static data. According to the similarity in the same group and the difference between different groups, transit authorities can formulate targeted measures for subway stations with similar characteristics. (2) An important assumption of traditional regression model is that the regression function for each point is the same. However, in reality, regression functions may not be homogenous, particularly for spatial data. Peak-hour passenger flow is a typical kind of spatial data collected in different stations. We employ the GWR model to determine the correlation effect between peak-hour passenger flow and land-use density. This model embeds the spatial heterogeneity of the data into the regression parameters. The weighting factor is the distance among observations. In view of the spatial nonstationarity of passenger flow, GWR can compute different regression functions for each subway station and discover the spatial distribution of the correlation coefficients.

This work can be further strengthened in the following aspects. The land-use density is only the quantitative information without floor area ratio. We can enrich the model by incorporating additional land-use data. From the perspective of algorithm improvements, various sequential clustering approaches should be implemented, and their performances should be compared. In addition, the GWR model can be integrated with additional time variables.

Acknowledgment This work is partly supported by the National Natural Science Foundation of China (51408019, U1564212, and 71503018), Beijing Nova Program (z151100000315048).

References

1. L. Sun, Y. Lu, J.G. Jin, D.H. Lee, K.W. Axhausen, Y. Lu, K.W. Axhausen, An integrated Bayesian approach for passenger flow assignment in metro networks. Transp. Res. C **52**, 116–131 (2015)
2. W.-T. Zhou, B.-M. Han, Passenger flow assignment model of subway networks under train capacity constraint. J. South China Univ. Technol. **43**(8), 126–134 (2015)
3. J. Wang, J.-F. Liu, F.-L. Sun, Passenger demand distribution and increasing trend over Beijing rail transit. Urban Transp. China **10**(2), 26–32 (2012)
4. X.-Y. Ma, A. Jin, M.-M. Liu, et al., Rail transit passenger flow characteristics in Guangzhou. Urban Transp. China **06**, 35–42 (2013)
5. J.-F. Liu, M. Luo, Y.-L. Ma, et al., Analysis on the passenger flow characteristics of Beijing urban rail network. Urban Rapid Rail Transit **25**(05), 27–32 (2012)
6. V. Aguiléra, S. Allio, V. Benezech, F. Combes, C. Milion, Using cell phone data to measure quality of service and passenger flows of Paris transit system. Transp. Res. C **43**, 198–211 (2013)
7. S. Srinivasan, Linking land use and transportation in a rapidly urbanizing context: A study in Delhi, India. Transportation **32**(1), 87–104 (2005)
8. J. Dill, Transit use at transit-oriented developments in Portland, Oregon, area. Transp. Res. Rec. **2063**, 159–167 (2008)
9. N. Zhang, X.-F. Ye, L. Jian-Feng, The impact of land use on demand of urban rail transit. Urban Transp. China **08**(03), 23–27 (2010)
10. S.-S. Peng, X.-P. Wu, S. Mei, Study on coordination between urban rail transit and land use based on GIS. J. Railw. Eng. Soc. **01**, 76–79 (2011)
11. A.K. Jain, Data clustering: 50 years beyond K-means, in *European Conference on Machine Learning and Knowledge Discovery in Databases* (2008), pp. 651–666
12. S. Aghabozorgi, A.S. Shirkhorshidi, T.Y. Wah, Time-series clustering – A decade review. Inf. Syst. **53**, 16–38 (2015)
13. T. Warren Liao, *Clustering of Time Series Data—A Survey* (Elsevier, New York, 2005)
14. X. Golay, S. Kollias, G. Stoll, D. Meier, A. Valavanis, P. Boesiger, A new correlation-based fuzzy logic clustering algorithm for fmri. Magn. Reson. Med. **40**(2), 249–260 (1998)
15. C.S. Möllerlevet, F. Klawonn, K.H. Cho, O. Wolkenhauer, *Fuzzy Clustering of Short Time-Series and Unevenly Distributed Sampling Points*, vol 2810 (Springer, Heidelberg, 2003), pp. 330–340
16. M. Kumar, J. Woo, J. Woo, Clustering seasonality patterns in the presence of errors, in *Eighth ACM SIGKDD International Conference on Knowledge Discovery and Data Mining* (ACM, 2002), pp. 557–563
17. K. Košmelj, V. Batagelj, Cross-sectional approach for clustering time varying data. J. Classif. **7**(1), 99–109 (1990)
18. T.W. Liao, B. Bolt, J. Forester, E. Hailman, C. Hansen, R.C. Kaste, J. O'May, Understanding and projecting the battle state, in *23rd Army Science Conference*, Orlando, FL, vol. 25 (2002)

19. R.H. Shumway, Time-frequency clustering and discriminant analysis. Stat. Probab. Lett. **63**(3), 307–314 (2003)
20. X. Wang, K. Smith, R. Hyndman, Characteristic-based clustering for time series data. Data Min. Knowl. Disc. **13**(3), 335–364 (2006)
21. A.M. Alonso, J.R. Berrendero, A. Hernández, A. Justel, Time series clustering based on forecast densities. Comput. Stat. Data Anal. **51**(2), 762–776 (2008)
22. A.K. Jain, Data clustering: 50 years beyond K-means, in *Joint European Conference on Machine Learning and Knowledge Discovery in Databases*, vol. 31 (Springer, Berlin, 2008), pp. 3–4
23. C. Genolini, X. Alacoque, M. Sentenac, C. Arnaud, Kml and kml3d: R packages to cluster longitudinal data. J. Stat. Softw. **65**, 1–34 (2015)
24. C. Genolini, B. Falissard, KmL: a package to cluster longitudinal data. Comput. Methods Prog. Biomed. **104**, e112–e121 (2011)
25. Z.-X. Tao, N. Zhang, B. Du, Research on the time &space distribution of in urban rail transport. Urban Public Transport **2**, 33–35 (2004)
26. J. Zhao, W. Deng, Y. Song, Y. Zhu, What influences metro station ridership in China? Insights from Nanjing. Cities **35**(4), 114–124 (2013)
27. M.G. Mcnally, The four step model, in *Handbook of Transport Modelling*, (Elsevier, Amsterdam, 2008), pp. 35–52
28. O.D. Cardozo, J.C. García-Palomares, J. Gutiérrez, Application of geographically weighted regression to the direct forecasting of transit ridership at station-level. Appl. Geogr. **34**(4), 548–558 (2012)
29. D.P. Mcmillen, Geographically weighted regression: the analysis of spatially varying relationships. Am. J. Agric. Econ. **86**, 554–556 (2004)
30. W.-Z. Pei, The basic theoretic and application research on geographically weighted regression, Ph.D. Dissertation, Tongji University, Shanghai, China, 2007

Chapter 7
An Initial Evaluation of the Impact of Location Obfuscation Mechanisms on Geospatial Analysis

Pedro Wightman and Mayra Zurbarán

7.1 Introduction

In these past few decades, the rise of mobile technology and GPS capable devices favored the demand of software for providing context related services and specifically location-based services (LBSs). In the scientific community, this has led to multiple scopes of research, especially concerning user's privacy. There have been plentiful developments of multiple mechanisms for protecting location information, that is, any information that may lead to the identification of the spatial surroundings of a subject. Some of these mechanisms are based on location obfuscation, which is explained as "the means of deliberately degrading the quality of information about an individual's location in order to protect that individual's location privacy" by Duckham and Kulik [1]. Nevertheless, there is a latent curiosity of whether or not the resulting degraded location information could still be used to perform geospatial analysis, which at the end is one of the main issues regarding the studies of georeferenced data.

It is true that the scientific community cannot and should not neglect the exploitation of user's digital footprints while these are made available freely and publicly. However, there are privacy concerns on the side of users and these may impact their willingness to share their personal data especially during a prolonged use of a service [2], besides specific privacy regulations that must be complied with based on the country where the data is originated. Volunteered geographic information (VGI) and crowdsourced data, as they provide information that can lead to the identification of users, their daily habits, and private preferences, should be protected. Xu and Gupta [2] claim also that user's privacy concerns translate into

P. Wightman (✉) · M. Zurbarán
Department of Systems Engineering, Universidad del Norte, Barranquilla, Atlántico, Colombia
e-mail: pwightman@uninorte.edu.co; mzurbaran@uninorte.edu.co

© Springer International Publishing AG, part of Springer Nature 2019
S. V. Ukkusuri, C. Yang (eds.), *Transportation Analytics in the Era of Big Data*,
Complex Networks and Dynamic Systems 4,
https://doi.org/10.1007/978-3-319-75862-6_7

the fear of losing control of personal information and can be the cause of stress and anxiety. Therefore, the adoption of privacy measures can increase the perception of privacy and trust on the providers of a service, leading to more compliance from users to adopt an LBS.

In order to perform valid spatial data analysis with crowdsourced data, even when it has been protected by the implementation of a location privacy protection mechanism (LPPM), the geospatial data should remain useful, meaning that comparable results should be reached with the original data as well as with the protected one.

This chapter presents an introduction to current LPPMs, with a focus on noise-based location obfuscation techniques. Then, a methodology is presented to evaluate the impact of LPPMs using a real case study, in which geostatistical analysis is performed over collected georeferenced data from the Twitter streaming API during 5 months in the city of Milan, Italy. This would serve as an initial assessment of the impact on the results of geostatistical inferences over obfuscated data at a city level scale and the identification of possible minimum levels of protection for user's location that will not interfere dramatically with the conclusion of the analyses.

7.2 Location-Based Services

LBSs have been a very active research topic in the last decade, thanks to the increasing use of electronic devices capable of calculating their location using technologies like GPS, Wi-Fi, cellphone antennas, beacons, and the mixture of them and, on the other hand, also the increasing number of applications that use this information, either as part of the business core or as a way to gather detailed data about their users.

Despite the fact that before the appearance of smartphones, some LBSs already existed, they worked based on static subscription of the area of interest, so they could not adapt their content to the actual location of the user. This change allowed service providers to track their users, and use this information to provide a better product. In addition, having all these data about individuals and their social circles, which revealed the providers what their users were doing, when, where, and with whom, permitted the reconstruction of their routines, and an insight on their personalities, their preferences, and consequently, their value to the market. The age of user surveying is living its golden age.

LBSs are constituted by a client–server architecture as defined in [3]. Within Location Services are found actors that perform specific roles that make possible the use of the service, like in Fig. 7.1, and these roles may be as follows:

- User: the LBS subscribed user who makes a request from a mobile device capable of obtaining the user's location.
- Server: the LBS server that processes the query and provides relevant location information requested by the user, such as points of interest (PoI) or navigation services.

Fig. 7.1 Location services
proposed architecture

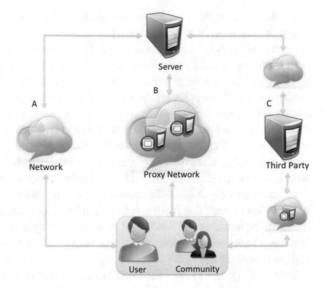

- Communication network: refers to a communication network such as the Internet, General Packet Radio Service (GPRS), or an ad-hoc network and any other means that make possible the communication between the user and the LBS server.
- Proxy: a service that provides security at network level to protect clients' location and identity through IP lookups, and these services could be distributed such as the onion router (TOR) [4] or centralized like virtual private networks (VPNs), the latter is not recommended for protecting specifically identity privacy for the arguments presented in [5].
- Community: denotes all the users of the LBS; the community may intervene in the functionality of the service, as is the case of applications used to monitor traffic. Community members could participate in methods for providing location privacy; however, not every LPPM requires a community to work.
- Third parties: are external relations that intervene to provide location privacy in conjunction with the LPPM. Third parties' relations act as proxy-like servers at application level that centralizes the architecture, in [6] it is defined as: "A subjective, dynamic, context-dependent, non-transitive, non-reflexive, non-monotone, and non-additive relation between a trustor and a trustee."

Given that people were not used to consider location as a sensitive information, compared to their social security number, they were willing to share it with everyone on every platform: social networks and their geolocated posts, apps that encouraged sharing their location (Foursquare, friend search, travel journals, geographical scavenger hunt, etc.), and apps that required location but was not part of their core functionality (video games, utilities, etc.). The main problem is that, especially in the first two platforms, this information is made public, so anyone could gather this information without any explicit authorization from the user, supported on what the terms and conditions of use had predetermined on the contract that all users accepted, but most likely did not read.

In recent years, all social network platforms are encouraging users to understand better the implication of their decision related to privacy policies, in order to try to limit the amount of information available to the public. Even though this is a positive trend, the service providers still have access to all the data, and there is no option for the user to opt out from the detailed acquisition of these data, including location; a promise of good faith on their service provider side is the only thing they have to trust the intentions of the companies that use this information for profit, especially when there is generalized lack of regulation specifically for dealing with location information. The European Union Directive on Privacy and Electronic Communications—2002/58/EC is one of the only pieces of legislation related to the information privacy that includes aspects of location privacy. In the USA, there is an ongoing initiative called S.2270—Location Privacy Protection Act of 2015 which was introduced to the senate, but never approved. Also, there was the SB 1434—California Location Privacy Act that in 2012 failed to become a state law after being vetoed by the governor, despite bipartisan approval on the state senate.

Location privacy can be defined as the ability of users to decide what location information they want to share, when, how, whom they want to share it with, and until when they prefer to do it [7]. Current LBSs do not offer but one option: "give us permission to obtain your location or do not use our app." Even if the user can decide if the location information is not shared with the public, the company still has full access to it. Users have no way to verify if their information is being used according to the contract, and if the aggregated information is really hiding their information and not allowing individualization, which is a privacy risk. In addition, if there is a security breach at the server side, the information can be stolen and used later against users.

The need to include user-centric LPPMs in all the services that individuals use should be a priority. Regulation should require the inclusion of mechanisms to actively protect the location information, allowing users to configure it as they consider adequate according to their preferences. An initiative on this direction could not be done only by ordering all service providers in the world to include them in their software, because it would be impossible to verify the real implementation of the tools, etc. Instead, it would require the participation of the manufactures of the devices and their operating systems; in order to guarantee that, despite the nature of the application, the location information provided internally by the phone to the app would comply with the privacy policies defined by the users.

In the next section, we will present an overview on the location privacy protection mechanisms in the literature that could be used to provide an extra layer of location privacy from the user side.

7.3   Location Privacy Protection Mechanisms

LPPMs comprehend the methods developed to provide location privacy; these vary in the requirements for implementation, processing and hardware resources, architecture, privacy provided, and the applications—location services—that are supported. In the following section, a taxonomy and a brief description of the techniques are provided.

7.3.1   Taxonomy of Location Privacy Protection Mechanisms

In the literature, there are many different flavors of protection mechanisms, depending on the application. In general, a basic differentiation can be done between two types of services: proactive services, in which the user's application is constantly sending the location in order to be registered in the system, i.e., tracking, geofencing and children security, traffic, etc., and reactive, in which the user triggers the service by sending his location and a specific request, usually related to a PoI, i.e., social network posts, geolocated search for restaurants, tourist attractions, people, etc. [8].

Each of these services poses different challenges to protecting the privacy of its data, thus different techniques have advantages that are adequate for each scenario. Figure 7.2 shows a proposed taxonomy to classify the most common types of techniques; however, some of them can be used on both types of services, depending on the approach taken by the specific mechanism [9]. A brief description of each of the types is presented:

- Cryptography-based methods offer secure communication and preserve location information accuracy. These methods include the use of symmetric encryption, hash, and other types of transformation of the data, in such a way that just a reduced group of trusted users can calculate the original information. The main problem with these techniques is that, unless they use homomorphic encryption techniques, an untrusted service provider would be oblivious to the user's location, serving only as a blind proxy.
- Location obfuscation mechanisms and specifically noise-based LPPM transform the user's location in a way that the original location is permanently lost by adding randomly generated noise; however, the resulting obfuscated location is still close enough to be used by an LBS and provides an acceptable quality level.
- Pseudonyms are an alternative to provide identity privacy in location-based applications; however, the use of pseudonyms alone is not sufficient to provide location privacy in an LPPM, given that if a pseudonym stays the same over time, it will eventually end up with the identification of a unique user. Also, if the location is not altered, other sources of information available to an attacker can reveal all individuals related to a certain address.
- Private Information Retrieval allows the creation of a common language between the client and server applications, not known by an attacker. In this technique, the

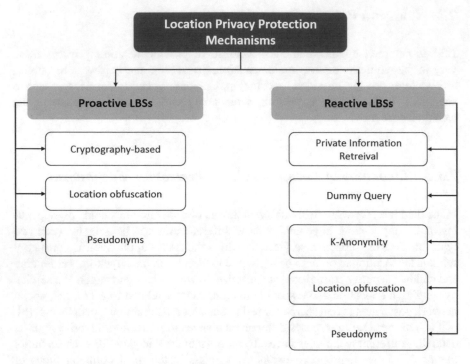

Fig. 7.2 Taxonomy of location privacy protection mechanisms (LPPMs)

data is coded in such a way that the server can work with the data as sent and offer all the available LBSs without actually revealing the original location of the user.

- Dummy query techniques consist on sending N fake requests along with the real one in order to disguise the user's true location. This technique poses downsides as it requires the server to process N additional queries, creating computing and communication overhead; however, there are some techniques developed based on dummy queries that manage to decrease such costs, such as piggybacking or using special coding to allow redundancy in search areas.

- K-anonymity was one of the first approaches presented to achieve location privacy; it consists on making a user indistinguishable among other $K-1$ users. In some implementations, the user may be able to specify the k parameter. In many of these techniques are used cloaked regions to provide such anonymity, where k users are similar enough within an area; however, regions with higher density of users result in smaller cloaked areas, or areas with very little population may require very large cloaked areas that reveal the existence of an outlier. Both cases derive in providing little protection. In general, a large of number of techniques in this category require the existence of a trusted third party that will have the necessary information to calculate either the K parameter or the best cloaking area, or the techniques may assume direct communication among users, which also presents technical challenges as well as issues related to security.

- Methods based on Progressive Retrieval (PR) perform many requests for a single user interaction, having each one of them modifying the search area, in order to reach the objective without pinpointing the user's original location. This approach aims to reveal as least location information as possible to obtain the desired service performance.

For more details on these techniques, please consult the works [10–13]. In this chapter, location obfuscation mechanisms were selected for the evaluation, especially noise-based obfuscation, due to their simplicity to implement in a real environment (no third-party actors, special hardware, or changes on the server side are necessary) and its nonreversible nature, which will permanently alter the user's data and makes difficult to recover the original locations, and the fact that it can be parametrized by the user, so the definition of a default value is critical because most users may not be interested in changing it. Measuring the impact of one of the most likely mechanisms to be adopted by industry, can provide an initial insight on how much information is actually lost for common geostatistical studies, and which could be a good starting point for providing a minimum level of security to the users while maintaining a satisfactory level of accuracy.

7.3.2 Noise-Based Location Obfuscation

Noise-based location privacy techniques are one of the simplest ways to protect the exact location of the users. Its main characteristic is the induction of random noise to the original location obtained by any location provider (GPS, Cellphone tower, Wi-Fi, etc.) in order to alter it permanently. Due to the randomness of the noise, it is very difficult to recover the original location, which becomes a useful attribute in terms of security.

Other important advantages are: it does not require major changes on the server side, because the reported information still is a single valid geographical location, so it should not alter the data structure; it can be calculated on the device via simple calculations thus computational complexity is low; and it does not require a third-party element. In addition, the user can customize these techniques to define the desired amount of noise that will be applied to the locations; this range can go from 1 m to a few kilometers, depending on the application's need for accuracy, and the perception of security that users have about their decision.

A naive approach to these techniques would be that a larger noise area would be beneficial in terms of security; however, it will degrade the quality of service of some applications. One example of the impact of the amount of noise are geolocated marketing services: If a user has a noise level of 3 km, it may be reporting its location "nearby" a store in which he or she is not really close to; thus, a geolocated offer could be lost if it is a time restricted one.

Generating random noise is not a new concept; several statistical random distributions can generate the values needed for this: Gaussian, Uniform, etc. The noise generation algorithm is the core element in these techniques. Three techniques will be presented in this section, all of them based on a uniform distribution; this is because, it guarantees the highest variability on the data, which facilitates the generation of faraway points from the center compared to a Gaussian distribution.

7.3.3   The Rand Algorithm

Rand is the simplest among all noise-based location privacy techniques: define an open ball b (p, r) where p is the original location point and r is the maximum amount of noise that can be added to the location. Select a point from that open ball, p', and report that point as the user's location to the system [7].

The generation of a single point inside an open ball can be done in two different ways, as shown in Fig. 7.3. The first one is a Cartesian method, in which a random number is generated for each component of the coordinate, added up, and then the new point is verified to fall inside the circle because it could fall on the external areas on the square area of $2r \times 2r$.

The second one is the Polar method, in which a distance and an angle are generated, transformed into Cartesian coordinates, and added to the original point p. This technique does not require validation because the random distance is validated to be between 0 and r, this ensures that it cannot fall outside the circle.

Despite the fact that both generate a single point, the distribution of the random points, when run several times, shows differences between these techniques which may affect the impact of the algorithm. Figure 7.4 shows the graphical distribution of 500 points using the Cartesian and the polar approaches.

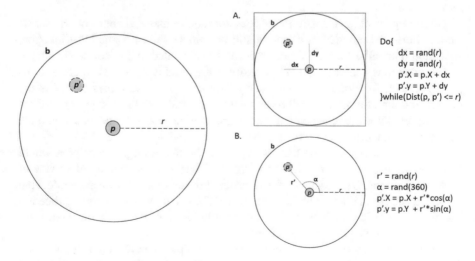

Fig. 7.3 Generation of a single random point

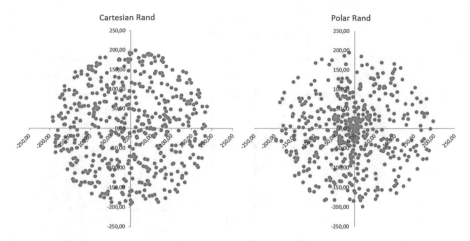

Fig. 7.4 Generation of 500 random points with Cartesian and Polar approaches using Rand

Distance Distribution Rand

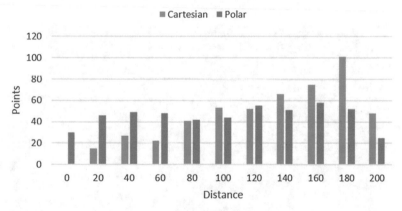

Fig. 7.5 Distance distribution of 500 random points with Cartesian and Polar approaches using Rand

Despite the fact that Cartesian-generated points look more evenly distributed along the area, without a concentration of points, this impression is false: Polar-generated points are distributed more evenly along the radius. The fact that there seems to be more points closer to the center shows that the distance among them is shorter, generating greater point density. Figures 7.5 and 7.6 confirm this fact. The distance distribution is clearly biased on the Cartesian approach, which tends to generate points farther from the center, while the Polar approach generates a more even amount of points per distance slot. The angle distribution is uniform in both approaches. Figure 7.7 shows a projection of the points as if the circle area was a rectangle. In that figure, it is easily seen how the Cartesian points tend to be farther

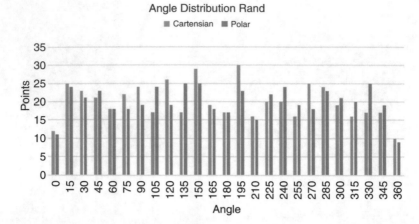

Fig. 7.6 Angle distribution of 500 random points with Cartesian and Polar approaches using Rand

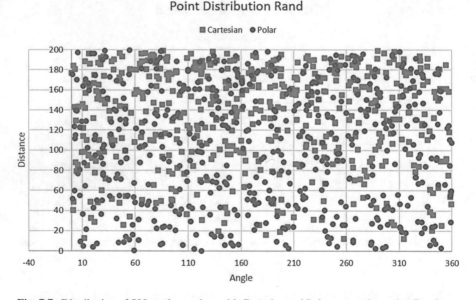

Fig. 7.7 Distribution of 500 random points with Cartesian and Polar approaches using Rand

away from the center. These differences may have an impact on the performance of the location privacy protection techniques.

The main advantage of this technique is that it is very simple to implement. However, it can generate values very close to the center, which is not entirely desirable because the protection for that specific point with no difference from the obfuscated one will be practically null.

7.3.4   The N-Rand Algorithm

N-Rand [10] is a modified version of the Rand algorithm in which N possible points are generated per location and just the farthest one to the original point is reported. This technique notably increased the average distance of the obfuscation but it decreased the variation of the noise; this fact increased its vulnerability to moving average-based filtering attacks to reduce the noise. In the paper, the authors showed that $N = 4$ presented a good balance between distance and variation. The original version was tested with Cartesian generation, but in this work it will be compared with Polar generation.

As it can be seen in Fig. 7.8, Cartesian generation generates a clear belt that starts around 150 m from the center location; this confirms the conclusions on reduction of variation showed in the original work. Polar generation shows a similar trend to drive points away from the center but in a softer manner. The distance distribution of the points shown in Fig. 7.9 shows how the right tail of the distribution is longer for the Polar generation, while the angle distribution behaves very similar to the one in Rand. Figure 7.10 shows the projection of the point distribution for N-Rand.

The distance distribution of the Polar generation should imply that the variation is higher among the points, which has a potential to generate points in such a way that it may be more resistant to filter-based attacks due to the increase in variance of the points.

One drawback of the Rand and N-Rand family of techniques is that both are fully symmetrical, which can be a problem against noise filtering attacks, like the TIS-BAD algorithm [14]. This technique uses a modified Exponential Moving Average to reduce the noise and estimate the original path of the user. The next technique, Pinwheel, can generate an asymmetrical dominion for the random point selection.

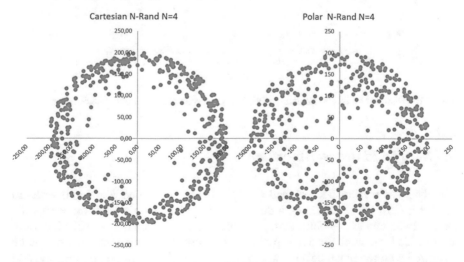

Fig. 7.8 Generation of 500 random points with Cartesian and Polar approaches using N-Rand $N = 4$

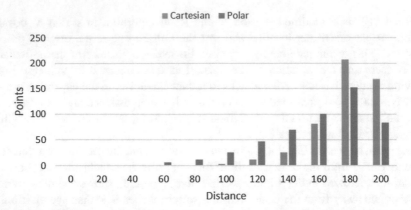

Fig. 7.9 Distance distribution of 500 random points with Cartesian and Polar approaches using N-Rand $N = 4$

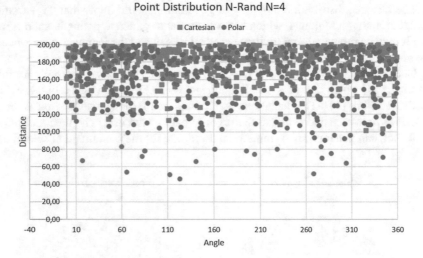

Fig. 7.10 Distribution of 500 random points with Cartesian and Polar approaches using N-Rand $N = 4$

7.3.5 The Pinwheel Algorithm

Pinwheel is a mechanism designed to add noise to the original location in order to distort it [15], hence providing a dummy or fake location that still represents the original and due to this blurriness, it is able to provide location privacy. The more noise added, the more privacy is provided, but also the location information can be distorted with too much noise in a way that may not be useful to provide any kind

Fig. 7.11 Graphical example
of the Pinwheel algorithm

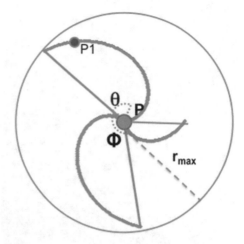

of service. The aim is to use a noise value that decreases the accuracy but that still provides useful context information about the location of the user.

Figure 7.11 presented a graphical representation of how the Pinwheel algorithm works, where P is the original location point at center of the circumference, r-max defines the radius, and φ defines the period for the repetition of the vanes that form the pinwheel. Each of which at a given α outputs the corresponding radius.

The noise added is calculated by defining a maximum radius, which is the maximum acceptable noise induced—distance from the original location, forming a circumference centered in it. The maximum radius serves to limit the dominium of the resulting obfuscated location, but this dominium is also determined by the pinwheel formula in Eq. (7.1); given that a θ value defines a specific radius for each θ within the circle described by the circumference, making the selection of the radius a deterministic process in the generation of random points with low cost of processing for doing on the fly transformations even on mobile devices, the formula used to calculate $r(\theta)$ is intended to be applied using polar coordinates and is as follows:

$$r\left(\theta\right) = \frac{\theta \bmod \varphi}{\varphi} + r_{\max} \tag{7.1}$$

The φ value has a great impact on how the random points are distributed. Figures 7.12 and 7.13 show the dominium of the random point generator with $\varphi = 12°$ and $\varphi = 110°$, respectively. In the first case, high periodicity and a symmetric distribution shows a distribution very similar to a uniform one, with polar generation approach. The second case shows a much lower periodicity and a very restricted area of point generation, which is also asymmetrical, increasing the probability of generating point in one sector of the circumference, which was shown to perform better against filtering attacks.

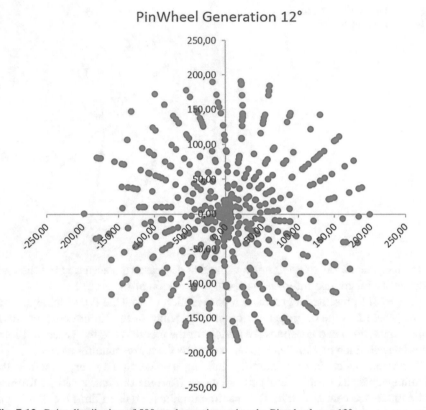

Fig. 7.12 Point distribution of 500 random points using the Pinwheel $\varphi = 12°$

Figure 7.14 shows the distance distribution of the first configuration compared with N-Rand. This was only calculated with polar generation given that generating Cartesian points to fit in the Pinwheel geometry would include unnecessary computational complexity to the algorithm. In general, both configurations showed a uniform distribution of the points, which is forced by the generation formula. For this reason, just one figure is shown.

Figures 7.15 and 7.16 show the square projection of the points which, as mentioned before, exemplifies the impact of the parameter φ on the definition of the domain of random point generation.

With noise-based techniques, once the user location is altered, the original data is lost. This means that what the service provider will receive will be only a probabilistic hint of where the user was, which, depending on the maximum radius, will have value for spatial data analysis used in user routine, mobility, or land-use studies.

Given that the implementation of a standard location privacy mechanism may happen soon, specially by organizations like the European Union that are very protective of the user's rights, it is critical to understand the impact on common geographical analysis so there would be an appropriate criterion to design this privacy layer for the general case, and it can be adopted easily by the industry.

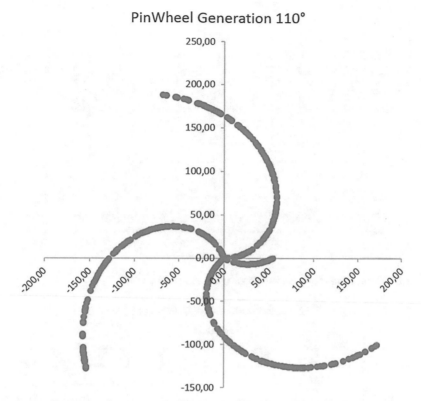

Fig. 7.13 Point distribution of 500 random points using the Pinwheel $\varphi = 110°$

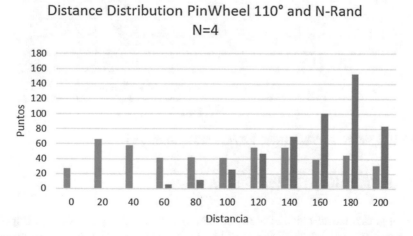

Fig. 7.14 Distance distribution of 500 random points using Pinwheel $\varphi = 12°$

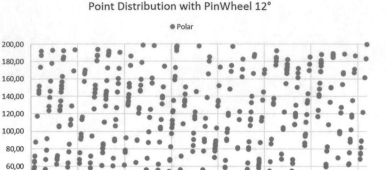

Fig. 7.15 Distribution of 500 random points using Pinwheel $\varphi = 12°$

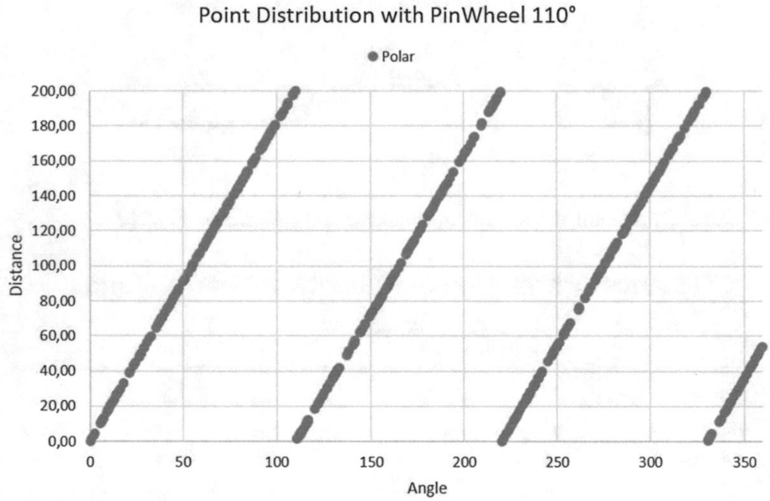

Point Distribution with PinWheel 110°

● Polar

Fig. 7.16 Distribution of 500 random points using Pinwheel $\varphi = 110°$

7.4 Exploratory Spatial Data Analysis

From a geostatistical point of view, randomness is not desirable, since it impedes inferences on spatial data, Griffith in [16] refers to Tobler's first law of geography: "everything is related to everything else, but near things are more related than distant things." Parting from this, Griffith introduces spatial autocorrelation as "the correlation among values of a single variable strictly attributable to their relatively close locational positions on a two-dimensional surface." Going further, Griffith states a dependency between values of a variable in neighboring locations due to underlying common factors. With this understanding, statistical analysis of data alone without considering their geographical context misses out on the possibility to make this correlation analysis.

Exploratory spatial data analysis (ESDA) aims to perform analysis of data when spatial autocorrelation matters, which is the case of most analyses when there is the possibility of acquiring the geographic component. ESDA includes different statistical techniques to describe and visualize spatial distributions and the identification of local patterns of spatial autocorrelation. Other kinds of exploratory analysis, like exploratory data analysis (EDA), proved useful when applied to nonspatial datasets to get insights before performing hypothesis tests and formal modeling but were shorthanded when ignoring the possibility of autocorrelation of the two-dimensional spatial variable, leading to incorrect assumptions of the independence of observations and therefore disregarding Tobler's first law of geography.

Currently, the use of ESDA is largely accepted as the best practice for the analysis of this kind of data, providing dynamic and interactive approaches allowing instant manipulation and visualization capabilities [17].

Analyses such as Kernel density maps, Hotspot analysis, and Voronoi maps provide dynamic spatial statistics following the principles depicted on ESDA.

7.4.1 Kernel Density Maps: Heatmaps

The use of Kernel density estimation on spatial data serves to identify the density of points in a neighborhood centered on each feature, providing local statistics comprising the features within a fixed search radius or bandwidth that are represented as clusters.

These classification maps, also referred as heatmaps, aid in exploring large datasets to identify atypical concentrations. For the means of visualization, the density values are classified within ranges and a color gradient palette represents these ranges. This classification may be subjective and strongly affect the resulting raster. Therefore, density maps alone cannot determine statistical significance.

To determine the bandwidth used in heatmaps is calculated the mean center of all input points and then this is used to determine the relative distance for each point, its medians, and standard distances. The following formula explains the calculated radius:

$$r = 0.9 * \min \left(\text{SD}, \frac{\sqrt{(1)}}{\ln(2)} * \text{Dm} \right) * n^{-0.2} \qquad (7.2)$$

where SD is the standard distance, Dm the median, and n is the total count of points in the dataset. The output map is a smoothed surface from overlapped circular density areas of each point [18].

7.4.2   Hotspot Analysis Using Moran's Index

Hotspot analysis uses the Getis-Ord local statistic Gi* [19], to define the areas of high point density occurrence or hotspots versus areas of low occurrence or cold spots. To perform a hotspot analysis, it is required to aggregate points; for this, the authors used the Hotspot Analysis Plugin for QGIS, which uses inputs that have the same coordinates and creates unique weighted values from them.

For the hotspot analysis are calculated the z-score and p values having as null hypothesis the assumption of complete spatial randomness. The Moran's Index is used to determine a bandwidth where the autocorrelation is higher, and this is done by maximizing the z-scores while iterating through different bandwidths, and then, the obtained optimal bandwidth is used to compare each point with its neighbors to obtain local Gi* statistics. Hotspot analysis in contrast with the density analysis is reliable to determine statistical significance.

To perform this analysis, it is necessary to create weighted points or features; each feature and its weight are the input for the algorithm to determine if they are hotspots, cold spots, or not significant. Usually, some kind of geographical clustering is used to create the features.

7.5   Methodology

The evaluation of the impact of LPPMs on spatial data analysis is a critical step before a massive adoption can take place. The result of this exploratory study will give insights on the impact of noise-based location obfuscation techniques in spatial analysis, and identify a preliminary level of noise in which the impact is minimum and provides a certain level of protection. Defining a general default noise value that guarantees protection and reduces interference is still an open problem in the field and subject to future research.

Each spatial data analysis technique has its own set up, depending on the resultant data:

- Hotspot analysis: the area of study is divided into regions, either political divisions or purely geographical. The original algorithm identifies areas with higher than normal data points (hotspots), or lower than usual (cold spots). After applying an LPPM, the maps are compared to calculate the number of false positives and false negatives.
- Heatmap: the map is colored based on the number of points associated to the hottest areas in the map. This number is expected to change when noise is applied, due to the diffusion of the points in the map.

The original dataset included a total amount of points of 18,966 from the city of Milan, Italy, from February to May of 2017, consisting of geolocated tweets. The comparison will show both the results with the original scenario and the results after the points were obfuscated using the Pinwheel algorithm previously described, with two different maximum radiuses: 500 m and 1 km. The period parameter φ was set to 105° on all the experiments in order to have lower periodicity and asymmetrical random point generation domain.

The first set of experiments include the analysis of distortion using the full dataset, for both heatmap and hotspot analysis. The second experiment presents a time lapse analysis on three moments of a single day, to show the possible impact on mobility pattern analysis. The experiments were performed on QGIS 2.18.

7.6   Analysis of Experiments (the original one, Pinwheel with maximum radius set to 500 m and to 1 km)

For the experiments, the three datasets were used to create kernel density maps using 1 km for the bandwidth and hotspot analysis where the Moran's I of the data was used to calculate the bandwidth value. Hotspot analysis was performed using the QGIS plugin presented in [6]. In order to create weighted points with the geolocated tweets, a vector layer of 85 neighborhoods of Milan was used to aggregate; the resulting features for the analysis are the polygon centroids with the corresponding count of tweets within each neighborhood.

In Figs. 7.17, 7.18, and 7.19, there is a major concentration depicted in strong orange in the center of Milan and the overall perception of the maps is similar; however, in the legends it is evidenced how the concentration is affected when noise

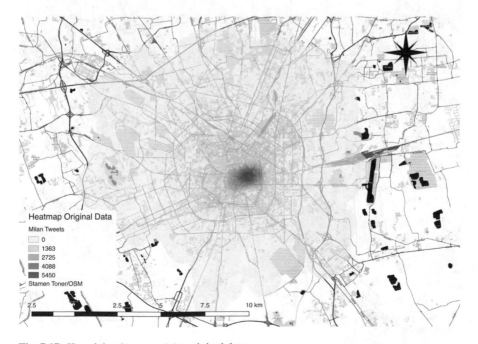

Fig. 7.17 Kernel density map of the original data

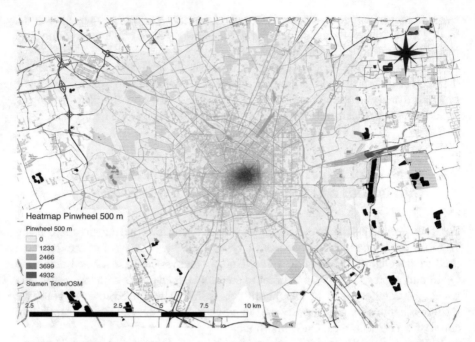

Fig. 7.18 Kernel density map of the obfuscated data with the Pinwheel algorithm set to 500 m of maximum radius

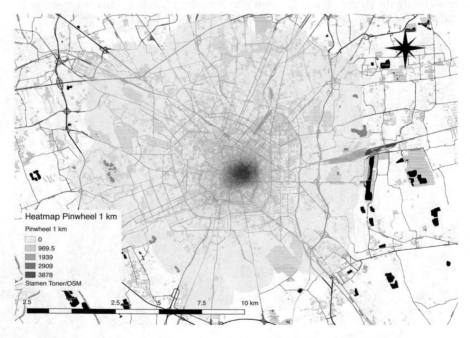

Fig. 7.19 Kernel density map of the obfuscated data with the Pinwheel algorithm set to 1 km of maximum radius

Fig. 7.20 Hotspot analysis map of the original Twitter data

is added. Figure 7.18 with 500 m of maximum noise has the highest value of density of 4932 and Fig. 7.19 with 1 km of noise has this value set to 3878, while with the original data of this value is of 5450. This variation exposes how the scale is altered; when noise is added, the density of the clusters becomes smaller. This is due to the elimination of overlapping points and more unique locations introduced by the randomness of the Pinwheel algorithm. Nevertheless, the density relation in the city remains, allowing to identify the same clusters of abnormal concentrations even with the obfuscated datasets.

To evaluate less subjectively the identified cluster, it was performed a hotspot analysis. Figure 7.20 shows the original hotspot analysis map and Figs. 7.21 and 7.22 show the output of the analysis for the obfuscated data with 500 m and 1 km of maximum noise, respectively.

There were no cold spots identified in any of the experiments and the Moran's I resulted in a value of 1200 m for the three datasets. The color of each feature (aggregated tweets in neighborhood centroids), represents the confidence level of their probability in respect to the null hypothesis (complete randomness), and the original analysis in Fig. 7.20 shows a concentration of four neighborhoods marked as hotspots in the center of the city.

On the output maps for the obfuscated datasets, there are still marked as hotspots the four neighborhoods that appear with the original dataset, but with 500 m of maximum noise a new one is marked as a 90% confidence hotspot. With the dataset

Fig. 7.21 Hotspot analysis map of the obfuscated data with the Pinwheel algorithm set to 500 m of maximum radius

Fig. 7.22 Hotspot analysis map of the obfuscated data with the Pinwheel algorithm set to 1 km of maximum radius

obfuscated with 1 km of maximum noise, there are two new hotspots marked in contrast with the original data, one with 99% of confidence and the other with 90% of confidence.

This is explained due to the alteration of the reported original tweets with the Pinwheel algorithm; the density as seen in the kernel maps decreases, making the distribution more spread. The count of tweets within neighborhoods tends to become more even, increasing the possibility of new hotspots where the concentration was higher before adding noise. The maximum radius set to the Pinwheel algorithm; however; does not allow a drastic change within the city that has an area of approximately 15 km^2. In the experiments, there was a reduction on the points considered, since when noise is added the tweets that are placed in the outskirts of the city may result outside the city boundaries used for the analysis aggregation. When maximum radius was set to 500 m, 135 points were left out and when using 1 km, 196.

To understand critical clusters throughout the day, a weekday was selected from the dataset. Thursday 4th of May Twitter data was split into different time spans and three of them were selected: from 7 to 10 am, from 5 to 7 pm, and from 9 to 11 pm. These were selected in order to understand mobility within the city and Figs. 7.23, 7.24, and 7.25 depict the result for heatmaps using 400 m for the bandwidth value.

In the morning in Fig. 7.23, there are two important clusters in the center of the city and very few data around the outskirts. When people are assumed to be leaving work around 5 pm as seen in Fig. 7.24, there is wider spread of points and a third cluster can be identified, and this corresponds to the Sempione Park, which is an

Fig. 7.23 Heatmap of tweets in Milan on a weekday from 7 to 10 am

Fig. 7.24 Heatmap of tweets in Milan on a weekday from 5 to 7 pm

Fig. 7.25 Heatmap of tweets in Milan on a weekday from 9 to 11 pm

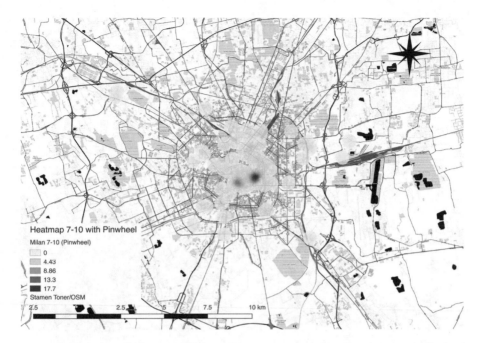

Fig. 7.26 Heatmap of obfuscated tweets in Milan on a weekday from 7 to 10 am with Pinwheel with a maximum radius of 500 m

important landmark of the city; and a higher concentration of tweets in the outskirts. From 9 to 11 pm, there is a new cluster, which corresponds to a dining event taking place in those days and more tweets reported from the outskirts and around a main station (Porta Garibaldi). This suggests a movement taking place from the center to outside of Milan.

To evaluate how similar studies may be affected when noise is added to the data, the three samples were obfuscated with Pinwheel with a maximum radius of 500 m. The results are seen in Figs. 7.26, 7.27, and 7.28.

The figures with the obfuscated datasets still provide an overall view of what was inferred with the original data but are more spread from what is seen in the legends with a decrease in the scales for all three experiments. In Fig. 7.28, from 9 to 11 pm the map is relatively more distorted not showing the particular cluster previously seen that corresponds to the dining event.

7.7 Conclusions and Open Problems

ESDA is oriented to determine valid inferences from geospatial data that are useful in many fields: mobility, land use, urban studies, etc. These inferences depend on the quality of the data obtained from individuals, thus any alteration of the data

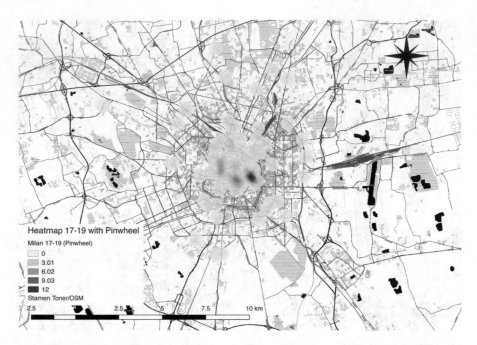

Fig. 7.27 Heatmap of obfuscated tweets in Milan on a weekday from 5 to 7 pm with Pinwheel with a maximum radius of 500 m

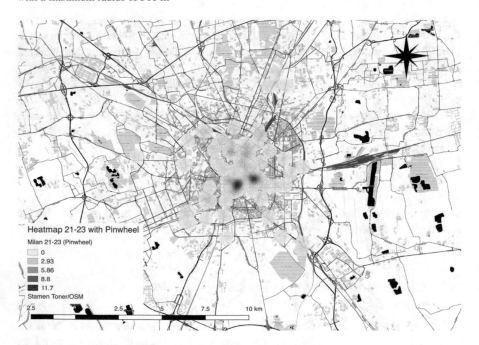

Fig. 7.28 Heatmap of obfuscated tweets in Milan on a weekday from 9 to 11 pm with Pinwheel with a maximum radius of 500 m

will definitely affect the results of any analysis and eventual decisions made using these results. On the other hand, citizens have the right to protect their personal information, including their location and specially when it is being used without their explicit consent.

If location privacy gains popularity in the policy agenda, it may become mandatory and its adoption will be imminent all around the globe. Professionals in ESDA must be prepared to deal with this information when it will produce probabilistic results, with some new kinds of false positives and false negatives that did not have to be expected in the past.

In the experiments, it can be noted that while in the kernel density maps the outputs are fairly similar, this is not the case with the hotspot analysis, where the maps are considerably altered with increased noise. From both analyses in the experiments, is concluded that, in order to reach equilibrium between the privacy protection and the quality of the inferences, a noise greater than 500 m strongly affects the results and should be avoided, for a study of a city scale, like Milan. This value provides an insight into the feasibility of implementing location privacy algorithms and a proposed default setting for them.

Also, it is worth to mention that these conclusions are valid for dense urban areas, where the amount of data is large enough to balance out the alteration of the points, compared to rural areas in which the population may be very small and a single change will alter the complete picture.

This chapter only focused on two types of geographical data analysis, but there are futher more available in the literature as well as location privacy protections mechanisms. Work needs to be done in terms of evaluation of different combinations of geospatial analyses and LPPMs.

Acknowledgments Acknowledgements to the Sustain-T Project (Technologies for Sustainable Development) by Erasmus Mundus partnership for supporting the author and encouraging international cooperation in research.

References

1. M. Duckham, L. Kulik, in *A Formal Model of Obfuscation and Negotiation for Location Privacy*, ed. By H.-W. Gellersen, R. Want, A. Schmidt. Pervasive Computing (Springer, Berlin, 2005), pp. 152–170
2. H. Xu, S. Gupta, The effects of privacy concerns and personal innovativeness on potential and experienced customers' adoption of location-based services. Electr. Mark. **19**, 137–149 (2009)
3. C.S. Jensen, H. Lu, M.L. Yiu, Location Privacy Techniques in Client-Server Architectures, in *Privacy in Location-Based Applications*, (Springer, Heidelberg, 2009), pp. 31–58
4. TOR project. Online: https://www.torproject.org. Last seen on May 18, 2017.
5. J. Appelbaum, M. Ray, K. Koscher, I. Finder, *vpwns: virtual pwned networks*. In 2nd Workshop on Free and Open Communications on the Internet, 2012
6. D. Oxoli, M.A. Zurbarán, S. Shaji, A.K. Muthusamy, Hotspot analysis: a first prototype Python plugin enabling exploratory spatial data analysis into QGIS. PeerJ Prepr. **4**, e2204v4 (2016)

 7. M. Duckham, L. Kulik, Location privacy and location-aware computing. Dyn. Mobile GIS **3**, 35–51 (2006)
 8. M.A. Labrador, A.J. Perez, P.M. Wightman, *Location-Based Information Systems: Developing Real-Time Tracking Applications* (CRC Press, Boca Raton, 2010)
 9. M. Zurbarán, L. González, P. Wightman, M. Labrador, A survey on privacy in location-based services. Ing. Desarrollo **32**(2), 314–343 (2014)
10. P. Wightman, W. Coronell, D. Jabba, M. Jimeno, M. Labrador, in Evaluation of Location Obfuscation Techniques for Privacy in Location Based Information Systems. *2011 IEEE Latin-American Conference on Communications (LATINCOM)* (2011), pp. 1–6
11. M. Zurbaran, K. Avila, P. Wightman, M. Fernandez, in Near-Rand: Noise-based location obfuscation based on random neighboring points. *2014 IEEE Latin-America Conference on Communications (LATINCOM)* (2014), pp. 1–6
12. C. Ardagna, M. Cremonini, S. De Capitani di Vimercati, P. Samarati, An obfuscation-based approach for protecting location privacy. IEEE Trans. Dependable Secure Comput. **8**, 13–27 (2011)
13. G. Ghinita, P. Kalnis, A. Khoshgozaran, C. Shahabi, K.-L. Tan, in Private Queries in Location Based Services: Anonymizers are not Necessary. *Proceedings of the 2008 ACM SIGMOD international conference on Management of data* (2008), pp. 121–132
14. M.A. Labrador, P. Wightman, A. Santander, D. Jabba, M. Jimeno, Tis-bad: a time series-based deobfuscation algorithm. Invest. Innovación Ing. **3**(1), 1–8 (2015)
15. P. Wightman, M. Zurbaran, A. Santander, in High Variability Geographical Obfuscation for Location Privacy. *2013 47th International Carnahan Conference on Security Technology (ICCST)* (2013), pp. 1–6
16. D.A. Griffith, Methods: Spatial Autocorrelation A2 - Kempf-Leonard, Kimberly, in *Encyclopedia of Social Measurement*, (Elsevier, New York, 2005), pp. 581–590
17. L. Anselin, in *Interactive Techniques and Exploratory Spatial Data Analysis*, ed. By P. Longley, M. Goodchild, D. Maguire, D. Rhind. Geographical Information Systems: Principles, Techniques, Management and Applications (Geoinformation Int., Cambridge, 1999)
18. B.W. Silverman, *Density Estimation for Statistics and Data Analysis* (CRC Press, Boca Raton, 1986)
19. A. Getis, J.K. Ord, The analysis of spatial association by use of distance statistics. Geogr. Anal. **24**, 189–206 (1992)

Chapter 8
PETRA: The PErsonal TRansport Advisor Platform and Services

Michele Berlingerio, Veli Bicer, Adi Botea, Stefano Braghin, Francesco Calabrese, Nuno Lopes, Riccardo Guidotti, Francesca Pratesi, and Andrea Sassi

8.1 Introduction

Smart Cities applications are fostering research in many fields including Computer Science and Engineering. Data Mining is used to support applications such as optimization of a public urban transit network [6], event detection [5], and many more. Along these lines, the aim of the PErsonal TRansport Advisor (PETRA) EU FP7 project[1] is to develop an integrated platform to supply urban travelers with

[1]http://www.petraproject.eu.

M. Berlingerio · A. Botea (✉) · S. Braghin
IBM Research Ireland, Dublin, Ireland
e-mail: mberling@ie.ibm.com; adibotea@ie.ibm.com; stefanob@ie.ibm.com

V. Bicer
Core Media, Dublin, Ireland
e-mail: velibicer@coremedia.ie

F. Calabrese
Vodafone, Milan, Italy
e-mail: francesco.calabrese@vodafone.com

N. Lopes
TopQuadrant, London, UK
e-mail: nlopes@topquadrant.com

R. Guidotti · F. Pratesi
KDDLab Department of Computer Science, University of Pisa, Pisa, Italy
e-mail: guidotti@di.unipi.it; pratesi@di.unipi.it

A. Sassi
Epoca, Modena, Italy

© Springer International Publishing AG, part of Springer Nature 2019
S. V. Ukkusuri, C. Yang (eds.), *Transportation Analytics in the Era of Big Data*,
Complex Networks and Dynamic Systems 4,
https://doi.org/10.1007/978-3-319-75862-6_8

smart journey and activity advices, on a multi-modal network, while taking into account uncertainty. Uncertainty is intrinsic in a transit network, and may come in different forms: delays in time of arrivals, impossibility to board a (full) bus, walking speed, as well as incidents, weather conditions, and so on. The PETRA consortium includes three cities, with different characteristics and problems, and different sources of uncertainty: Rome (Italy), Venice (Italy), and Haifa (Israel). While in Rome and Haifa the platform built within the project is enabling demand-adaptive journey planning intended mainly for commuters, PETRA's objective in Venice is enabling smarter tourism. This is an area where predicting the next locations visited by tourists [1], recommending new locations to visit [12], and, more in general, mining information related to tourism activities [26], has been key to better support crowds and individuals when they visit a location during their holiday or free time.

In the work, we describe the high level architecture of the PETRA platform, and present the results obtained by applying it to two different use cases coming from two of the three partnering cities: Rome and Venice.

In Rome, we applied the embedded Journey Planner (JP) on thousands of planning requests, performed with and without the results coming from the Mobility Mining module. We show how, by integrating private transport routines into a public transit network, it is possible to devise better advices, measured both in terms of number of requests satisfied and in terms of expected time of arrivals. These experiments are part of the validation for the PETRA use case on Rome, where we assess the quality of the advices coming from the innovative integrated platform. The specific PETRA use case for the city of Rome revolves around an enhanced JP, capable of leveraging information about other users to provide optimal solutions. Furthermore, connecting the JP to the Data Manager (DM) allows the JP to efficiently react to changes in the transportation network. Such changes can automatically be detected by city sensors or manually inserted by city transportation managers using the PETRA dashboard, a tool that interacts with the stored General Transit Feed Specification (GTFS) data,[2] allowing to directly interact with city transport information. This chapter describes two core components of the PETRA technology stack, the JP and the DM, and their application to Rome city's data.

In Venice, we couple big data fusion, management, and mining, to support smarter tourism. At a high level, given a set of desired Points of Interest (POIs) to visit in a city, we devise activity plans that take into account historical patterns of crowding level at a landmark, or along the path to reach it, along with the projected paths from the active tourists who are currently visiting the city. Our main goal is to order the sequence of POIs to visit, set an appropriate departure time, constantly monitor the crowding level of the city, and re-plan as appropriate, to reduce the impact of large crowds of people visiting a city. This is a real problem in the city of Venice: many tourists visiting the historical city centre (with narrow streets and, in general, a small pedestrian area) often lead to pedestrian congestion that the city needs to handle, typically by sending police forces to help restore the

---

[2]https://developers.google.com/transit/gtfs/reference.

normal pedestrian flows. The city of Venice has also instituted the profession of "intromettitore"[3] (i.e., "intruder"), whose task is to stop incoming tourists where they typically approach the city (the main train station, the main bus station, and the main ports for water taxis), listen to their visit preferences, and offer incentives (like free water taxi rides) to follow alternative plans, serving two purposes: reducing congestion caused by all the tourists wanting to see the most popular landmarks, and increasing the visibility of less popular places, with a better distribution of the impact on the local economy. Without offering incentives, and without suggesting different places to visit, our goal is to build a system serving the same two purposes, but in an optimized and automated fashion, by devising plans that help spreading the tourists over the entire pedestrian area of Venice, with the side effect of also making less popular areas more visible. To do so, we built *togetThere*, a system that: (1) mines historical sensed presence data related to tourism activities; (2) devises tourism activity plans taking into account historical patterns of crowding levels at landmarks and along the pedestrian paths; (3) projects the active plans to better estimate the crowding levels for the rest of the day; (4) constantly monitors active plans and crowding levels for smarter re-planning and, finally, (5) collects user experience at the end of the visits to improve the planning algorithm. Our system has been designed, built, and tested (through simulations) on real data coming from the 2015 Telecom Italia BigData Challenge[4] and is now being deployed in the city of Venice within the context of the PETRA project.

8.2   Related Work

Work related to our study are in the areas of journey planning, smarter tourism, and data-driven mobility applications in general.

See [2] for a relatively recent survey on journey planning, including journey planning in road networks, and multi-modal journey planning.

Mamei and Zambonelli [23] discuss an approach to achieving load balancing inside a building such as a museum. The authors use a concept of a *crowd field*, a continuous mapping of the museum floor area into a field of "hills and valleys" that reflect the level of crowdedness in each room. The more crowded a room, the higher the "altitude" of the crowd field in that room. A hill-descending exploration strategy would encourage tourists to visit less crowded areas. The crowd field can be combined linearly with a mapping reflecting what rooms are more interesting to the tourist at hand, to obtain a strategy that allows to visit interesting locations while avoiding a high level of crowdedness.

---

[3]http://www.comune.venezia.it/flex/cm/pages/ServeBLOB.php/L/IT/IDPagina/2595.
[4]http://www.telecomitalia.com/tit/en/bigdatachallenge.html.

In outdoors tourist environments, performing tourist load balancing is motivated in part by environment protection and sustainability reasons.

A given configuration of tourists in a touristic area (i.e., the number of tourists at each relevant location in the area) can be evaluated using an objective function. Lin et al. [22] argue that the Variance and the Gini-Simpson Index, both used as objective functions in previous mathematical models for tourist load balancing, have their own limitations. The authors introduce an objective function called the tourism utility function (TUF), which assigns a lower satisfaction score when a visited spot is over-crowded (i.e., when the number of tourists exceeds a given capacity).

In designing a strategy to dispatching tourists through a given touristic area, several authors represent the area as a graph with relevant locations and links [14, 22, 31]. Dispatching strategies discussed in the literature include stochastic (probabilistic) approaches. For example, the average strategy (e.g., [22, 31]), applied at a given node in the graph, picks any of the successor nodes with an uniform probability. The space-time strategy (e.g., [22, 31]) favours nodes with a lower occupancy rate. Zheng et al. [31] combine the two strategies to obtain a reduction in the duration of bottlenecks (i.e., overcrowded spots).

Some authors [19, 21] have considered using Radio Frequency Identification (RFID) as a technology suitable for tracking tourists. This information can provide a view on the crowd levels at various locations within the monitored area, and it can be utilized as input data to various tourist dispatching strategies.

Hsieh and Li [18] investigate a problem where route planning is combined with visiting a number of attractions. They consider the pleasure of visiting a location as a function of the time of visit. For example, it is much more pleasant to visit a beach at the sunset time, as opposed to the hot time interval around the noon. In contrast, our focus in the Venice use case is on load balancing both on the travel map and at the points of interest.

Qiu et al. [27] study the spatio-temporal distributions of visitors in a Chinese tourist site to provide a navigation for balancing the load of visitors within the site. The authors simulated the visitor data by assuming the capacity of sites, roads, visitor walking and sightseeing time, among others, and quantified the loads of each visitor spot. To balance the loads, the authors build a mathematical model that minimizes the variance of loads in each spot. The authors do not investigate how to find a solution to this mathematical model. The paper also points out that vehicle scheduling can be used to balance the loads within a site. It formalizes a plan for vehicle scheduling and presents a list of steps for simulating the scheduling. Again, it does not provide a method to solve the scheduling problem in this context.

Berlingerio et al. [4] apply data mining to telco data, inferring frequent mobility patterns for telco users. Individual mobility patterns are aggregated to obtain common mobility patterns for larger groups of users. These are fed into an optimization engine used to design a public transportation network optimized to users' mobility needs observed in the data.

Berlingerio et al. [7, 16] extract mobility demand and personal interest from Twitter data to enable a more sociable carpooling service. They perform a multi-objective optimization of cars, drivers, and passengers, aimed at increasing the "enjoyability" of a car while minimizing the total number of cars needed to meet the demand.

8.3   Use Cases Overview

In PETRA we focused on two types of use cases: the A to B journey planning scenario, and the Point of Interest (POI) visiting (or, activity planning) scenario. The first one has been investigated in the city of Rome, the second in the city of Venice.

8.3.1   Journey Planning in Rome

Consider the following problem: you are in Rome, Italy, and you want to move from location A to location B, using Rome's multi-modal transit network. You want a solution with the following characteristics, at least:

- being able to specify A, B, and user preferences such as maximum walking time, which transport means to include or exclude, starting time, and so on
- capable of computing a *robust* solution that takes into account various types of uncertainties in the transit network (delays, missed follow-up connections due to delays, etc.), at planning time (so you do not have to re-query the journey planner for every change in the transit network)
- being able to fully exploit the potential of the multi-modal transit network, possibly including private means of transport, such as private cars (that may be available to offer a ride to passengers).

Ideally, a user would open a mobile application (front-end of the solution) and input the query, which would connect to the back-end planning engine, would present the result, and should keep tracking both the user and the transit network for changes in the original plan.

As real life presents troubles along the way, there may be changes in the network, or the user may deviate from the plan: if these changes are within a "safe zone" of events that are predictable (for example, by looking at the historical patterns of delays in transit), they could be taken into account at planning time, thus avoiding re-planning. In that case, the mobile application can cache a robust *contingent plan* which is not just a sequence of actions that would bring you from A to B but that would resemble more a *tree* of possibilities that can be predicted beforehand. If, on the other hand, the changes represent a major disruption, then this should be detected as an edge case outside the contingent plan, and re-planning is needed. The concept of uncertainty in PETRA is further described in Sect. 8.4.

It is well known that most of our mobility is predictable [28], due to recurring patterns, typically home-work commutes, work-supermarket, supermarket-school, etc. So, it is reasonable to assume that people drive most likely following almost always the same routes, at almost the same time. This information could be exploited and private cars may be seen as "virtual buses" available to extend the coverage of the public multi-modal transit network. Mobility Data Mining [6] can be used to capture this knowledge and use mobility patterns for improving the transit network.

8.3.2 Activity Planning in Venice

This use case was brought to us by the mobility agency in Venice, Italy, which is dealing with tourism mobility management on a daily basis. In particular, Venice sees an average of 60 thousands tourists per day (with peaks of 250 thousands), which is comparable to the number of inhabitants of the town (60 thousands). Moreover, Venice being an historical town with small streets (pedestrian only), it has the peculiarity that large crowd movements can sometimes cause pedestrian congestions that can only be solved by police intervention. This is particularly relevant in the areas close to the city's main visitor attractions such as Piazza San Marco and Ponte di Rialto. At the same time, other roads which are not on the main tourist routes are more rarely utilized. Figure 8.1 shows an image generated by the city, estimating the density of pedestrian flows on each road. As it can be seen, there are potentially many alternatives going through "green roads" to go between any two attractions in the cities, while mostly the shortest paths are used and so become congested ("red roads").

Clearly, cities would like to promote tourism to sustain their local economy, but are faced with the need to improve the tourists experience while also keeping the urban mobility flowing for their citizens. To deal with the above problem, we propose a solution for distributed tourism management by acting on individual travel plans to achieve a city-wide crowd level balance using BigData analytics and optimization. The solution is based on:

- understanding the use of the city and popularity of areas from a combination of data sources: mobile phone locations, mobile app usage, and social media
- issuing city-level optimized plans to avoid local crowding through the use of a mobile app for tourists
- monitoring crowding in real time from the combination of the above data
- offering plans passing through less crowded areas to tourists, while preserving visit preferences

We identify three main stakeholders to benefit from the proposed solutions:

- tourists, which will perceived less congestions and queuing times during their visits;
- municipalities, which will have a more distributed crowd movement, that can also be better managed during cases on emergencies or evacuations;

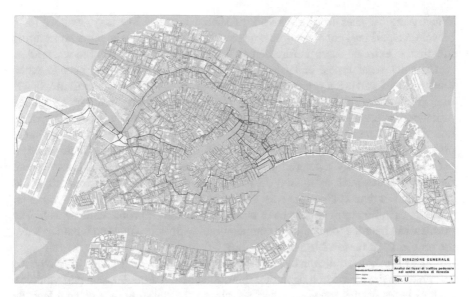

Fig. 8.1 Map of the city of Venice, with streets marked from green to red based on the estimated pedestrian flow

- local business activities, which will have a better perceived quality of service (due to less congestions) on the primary streets, and more visits on the secondary streets.

8.4   Modelling Uncertainty in Multi-Modal Journey Planning

In this section we overview a multi-modal journey planning system that is capable to reason about the uncertainty in the knowledge available in a multi-modal transportation network. As we illustrate in this section, this allows to compute journey plans that are less prone to failures due to unexpected events during a trip. This is a background section needed for a self-contained document. In the section we give references with more detailed information about the background information provided here.

A multi-modal transportation network can feature many types of uncertainty. The exact arrival and departure times of scheduled public transport vehicles, such as buses, can differ from the pre-planned, published schedules. This can further lead to missed connections, when a connection vehicle departs from a stop before another vehicle arrives at the same stop. The duration of legs such as driving legs, cycling legs and even walking legs can be non-deterministic. When a bus is too crowded, it may be impossible for new travelers to board the bus at a stop. When driving, waiting for a parking spot to become available could have an uncertain duration.

When using bicycles from a network of shared bicycles in a city, waiting for a bicycle or a parking spot to become available could also involve an uncertain waiting duration. In extreme cases, unforeseen events such as accidents can block a road.

Deciding what types, and how many types of uncertainty to encode when modeling and solving a problem is an important decision. At one extreme, deterministic problem modeling and solving ignores any uncertainty. The advantage of deterministic problem solving is that a problem instance can be smaller when uncertainty is ignored. This often translates into solving the problem faster. On the other hand, deterministic solutions may be too optimistic, reducing their practical usefulness. Informally, a deterministic solution, such as a journey plan, may ignore potential failures such as missed connections. Uncertainty-aware solutions could better avoid such risks along the way. See an example later in this section. At another extreme, encoding too many types of uncertainty could be problematic as well. The problem can become too large and computationally difficult, making it hard to solve the problem in a reasonable amount of time.

Our system can handle the following uncertainty related to the following types of information: the arrival and departure times of scheduled public transport vehicles, such as buses, trams, trains, or subway systems; the duration of a driving leg; the duration of a walking leg; the duration of a cycling leg; the waiting time until a parking spot becomes available at a car parking lot; the waiting time until a parking spot becomes available at a bike station; the waiting time until a bike becomes available at a bike station.

8.4.1 Uncertainty-Aware Journey Plans

A standard, deterministic journey plan is a totally ordered sequence of actions. In contrast, our uncertainty-aware journey plans can have a tree shape. The AI planning literature calls tree-shaped plans *contingent plans* [25]. Branching points in the plan provide more than one option at the execution time. If one option is not available, the traveler has the option on continuing on an alternative option, as shown in the example presented next. One can show that a contingent plan (i.e., a plan with branches) can be strictly more expressive than a sequential plan, thus being able to encode an optimal travel strategy in some cases where sequential plans lack this ability.

We use the examples shown in Figs. 8.2 and 8.3 to illustrate differences between sequential, deterministic plans and uncertainty-aware, contingent plans.

Figure 8.2 shows a toy transportation network where a user needs to travel from A to B. Observe the uncertainty in some of the arrival and the departure times. For example, at node C, the arrival time of the traveler is 9:30, plus or minus 5 min or less. There are two buses leaving from C to B. The first one departs at 9:35, plus or minus a few minutes, as shown in the figure. The second one departs at 10:35, give or take a few minutes.

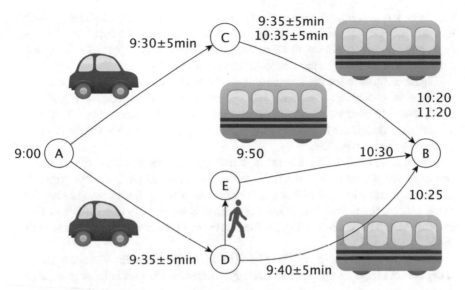

Fig. 8.2 Shortcomings of optimistic deterministic journey plans

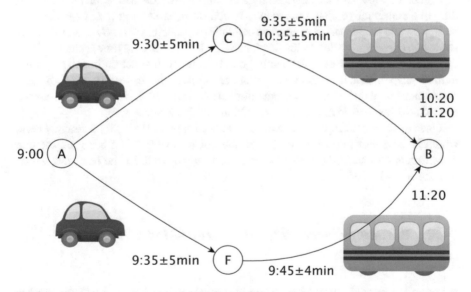

Fig. 8.3 Shortcomings of conservative deterministic journey plans

Due to such uncertainties, the traveler may catch or may miss the first bus. The limitation of deterministic journey planning is that a firm assumption needs to be made about catching the first bus. In other words, a deterministic journey planning algorithm has to commit to one of the following two options: it is definitely possible to catch the first bus, or it is definitely impossible to catch the first bus. When the first option is chosen, we say that the deterministic planner is optimistic. In the latter case, we say that the deterministic planner is conservative.

Figure 8.2 illustrates the shortcomings of optimistic deterministic plans. Shortcomings of conservative deterministic plans will be discussed in Fig. 8.3. An optimistic deterministic planner will recommend the trajectory A, C, B. The disadvantage is that, if the first bus is missed, the only alternative left for the traveler is to wait 1 h for the next bus.

On the other hand, in Fig. 8.2, the best strategy in terms of worst-case arrival time and expected arrival time is the following. First, drive to D. If the bus from D to B has not departed yet, take that bus. Otherwise, walk to E and take the bus from there. Indeed, the arrival time at the destination B varies between 10:25 and 10:30, as compared to the range from 10:20 (best case) to 11:20 (worst case) featured along the trajectory A, C, B.

Notice the if–then–else condition in D in the optimal strategy. A contingent plan is capable of encoding this optimal strategy, whereas a deterministic, sequential plan cannot.

Figure 8.3 illustrates the shortcomings of conservative deterministic plans. Recall that, in a conservative plan, uncertain connections are ruled out (i.e., considered to be impossible to catch). As such, the two distinct trajectories A, C, B and A, F, B show two conservative deterministic plans with the same arrival time, namely 11:20. Since the two trajectories look equally good in conservative deterministic planning, it may happen that a deterministic planner outputs the trajectory A, F, B as a recommended plan. However, the trajectory A, C, B is better, since there is a chance to catch the fast bus departing around 9:30 along this trajectory.

Once again, a contingent plan would be able to capture the optimal strategy here, which is the following: Drive to C and attempt to take the first bus (i.e., the one departing at around 9:35). If this bus is already gone, wait for the next bus, which departs around 10:35.

8.4.2 Encoding Uncertainty into a Multi-Modal Transport Network

We overview our definition of an uncertainty-aware *network snapshot*. The snapshot is a knowledge base with all the information available about a multi-modal transportation network. For more information, see [9].

The data included in a snapshot depends on the transportation modes available. The snapshot needs to include a set $\mathcal{L}$ of *relevant locations* on the map. Depending on the transportation modes available, the set of relevant locations can include stops for scheduled transport (e.g., bus stops), bike stations, car parking lots, and taxi ranks. In addition, given a user request to compute a journey plan, the origin and the destination are added to $\mathcal{L}$ as relevant locations, unless already present.

For scheduled public transport data, the snapshot includes route and trip information. $\mathcal{R}$ is a set of *routes*. A route r is an ordered sequence of $n(r)$ locations, corresponding to the stops along the route. Notice that in real life different routes can have an identical label displayed on a vehicle. For instance, two distinct buses can be labeled as route 4, but one can go from west to east and the other one can go from east to west. These are two distinct routes in our encoding, as two different (ordered) sequences of stops imply the existence of two different routes.

$\mathcal{T}$ is a collection of *trips*. Informally, each trip is one individual scheduled-transport vehicle (e.g., a bus) going along a route (e.g., the bus that starts on route 4 at 4:30 pm, going westbound). Formally, a trip i is a structure $\langle r, f_{i,1}, \ldots f_{i,n(r)} \rangle$, where r is the id of its route, and each $f_{i,k}$ is a probability distribution representing the estimated arrival time at the k-th stop along the route.

When private car sharing is allowed, we model trajectories of car trips available for sharing similarly to public transport trips. For instance, model a daily morning trip from the car owner's home address to their work similarly to a public transport trip, with a few stops along the way and timing information at each stop.

The snapshot can further include a table $\mathcal{W}$ of walking times for every pair of relevant locations. Each walking time can be modeled as a probability distribution. Similarly, a table $\mathcal{C}$ can provide the cycling time (as a probability distribution) between pairs of locations such as bike stations. Given a time of the day t and a car parking lot l, $\mathcal{T}_{\mathcal{P}}(l, t)$ is a probability distribution representing the waiting time until a parking spot is available, if the user arrives at time t. Similar probabilities are defined for the waiting time at bike stations, to get a bike or a parking spot for a bike.

The part of the snapshot structure encoding the public transport data is similar to the GTFS format.[5] GTFS, however, handles no stochastic data. Furthermore, our snapshots cover additional transport modes, such as shared-bike data.

The way the network snapshot is defined is partly justified by practical reasons. It encodes available knowledge about the current status and about the predicted evolution over a given time horizon (e.g., estimated bus arrival times). At one extreme, the snapshot can use static knowledge, such as static bus schedules, or static knowledge based on historical data. At the other extreme, increasingly available sensor data can allow to adjust the predicted bus arrival times frequently, in real time [10]. Our network snapshot definition is suitable for both types of scenarios, allowing to adjust the approach to the level of accuracy available in the input data.

---

[5]https://developers.google.com/transit/gtfs/reference.

8.4.3 DIJA: An Uncertainty-Aware Multi-Modal Journey Planner

The input to our planning system are a network snapshot, as described in the previous section, and a user query. The query states the origin, the destination, and the departure time. It further states so-called quotas, which are maximum allowed values for the number of legs in a trip, the walking time and the cycling time. It can contain flags about what transport means are allowed in the trip at hand.

Our planning system performs heuristic search. The search space is an and/or state space. We give an informal overview of the state space, followed by an informal description of our search strategy.

In describing the state space, we need to define states and transitions between states. The core components of a state s include a position p_s, a density function t_s,[6] and a vector of numeric variables q_s. Some auxiliary state components are used for correctness (e.g., to define what types of actions apply in which states) and for pruning [9]. In this section, we focus on the core components.

The position $p_s \in \mathcal{L} \cup \mathcal{T}$ represents the position of the user in the state at hand. The position can be either a relevant location on the map or a trip id, in which case the user is aboard that trip. The time t_s is a probability distribution representing the time when the user has reached position p_s. Quotas left in this state (e.g., for walking time, cycling time, number of legs) are available in q_s. In the initial state, the quotas are taken from the user query. They get updated correspondingly as the search progresses along an exploration path.

Actions (transitions) considered depend on the transport modes available. Examples include `TakeTrip`, `GetOffTrip`, `Cycle`, and `Walk`. `TakeTrip` actions can have up to two non-deterministic outcomes, success or failure. Along the success branch, the user was able to board the vehicle (because the user arrives at the stop before the departure of the vehicle). Along the fail branch, the user misses the connection because the vehicle departs before the user arrives at the stop.

When a `TakeTrip` action has two non-deterministic effects, each branch has a probability associated with it. The probability p of the success branch is computed as a function of two distribution probabilities: the one corresponding to the arrival of the user at the stop, and the one corresponding to the departure of the vehicle at the stop [9].

Other actions apart from `TakeTrip` always succeed and therefore they always are deterministic actions (i.e., applying such an action results in exactly one successor state).

In exploring the search space outlined earlier in this section, our system is based on the AO* algorithm. The algorithm requires many search enhancements to make

---

[6]For clarity, we stick with the case of continuous random variables. Discrete distributions are handled very similarly.

it scale to a domain such as multi-modal journey planning under uncertainty in a large city.

Our implemented enhancements can be grouped into two categories, heuristic functions and pruning techniques. A detailed and formal presentation is beyond the focus of this chapter. Instead, we provide an intuitive idea behind a few enhancements. Part of our DIJA's enhancements are presented in detail in previous work [9].

Heuristics guide the search into more promising areas of the state space. Given a metric such as the travel time, for instance, a heuristic function $h(s)$ is an estimation of the travel time from a state s to the destination. A heuristic function is *admissible* when it does not overestimate the actual value (i.e., the actual travel time in our example). Using admissible heuristics ensures that algorithms such as AO* produce optimal solutions.

We created admissible heuristic functions for several metrics: the travel time, the number of legs in a trip, the walking time, and the cycling time. The first two are used to guide the AO* search.

The second, the third, and the fourth are used for pruning. Recall that users can specify maximum acceptable amounts for the walking time, the cycling time, and the number of legs in a journey. A state s can be pruned (i.e., treated as a deadend) when the amount spent so far (for instance, the number of legs from the origin to s), plus a precomputed admissible estimate of the amount needed from here on (for instance, an admissible estimation of the legs from s to the destination), exceed the maximum acceptable amount (the max number of legs specified by the user).

When several bus stops are within walking distance from a current location, generating a walking action to each of them might be unnecessary (e.g., imagine that all stops are served by the same route). We have introduced rules for pruning part of the walking actions, and proved their correctness in the presence of uncertainty.

We also perform state-dominance pruning. Intuitively, a state s_1 dominates state s_2 if they have the same position p, and reaching p in state s_1 uses less time and quota amounts than reaching p in state s_2. In have proved that, in non-deterministic planning, the correct application of our dominance pruning depends on the types of the non-deterministic branches on the path to each of the two states considered, and implemented our dominance pruning strategy accordingly [9].

8.5   PETRA System Components

The various use cases defined within PETRA required the definition of a general purpose architecture for data management. Figure 8.4 shows a high level overview of the system. The entire architecture consists of a series of services that can be classified in three categories: data integration, data management, and application services.

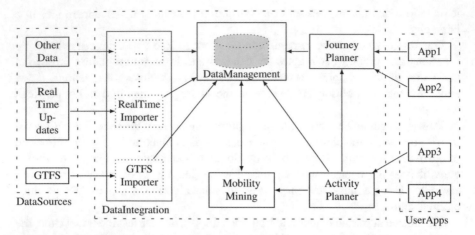

Fig. 8.4 High level PETRA architecture

8.5.1 Data Management

The PETRA project highlights the need to integrate different types of urban data, from unstructured data to real-time information retrieved from city sensors, structured public transport schedules and others. Handling large volumes of rich and heterogeneous urban data requires a tailored and scalable data management platform. For this purpose, PETRA leverages on Scalink. Scalink is a comprehensive data management platform developed within IBM Research Dublin that expands ElasticSearch[7] distributed data storage and indexing capabilities, providing indexes designed for the different formats of data chat can be handled by the system (relational, tabular, and graph data), and also their different types (geospatioal, textual, etc.). This is done extending ElasticSearch framework in such a way that it is able to handle natively LinkedData, thus extending ElasticSearch capabilities to seamlessly integrate heterogenous data. Furthermore Scalink expands ElasticSearch query and searching capabilities providing a combination of structural query processing and search techniques in order to satisfy the different kind of queries required by the various use cases.

At a high level, the data management module stores the data of each city as one massive RDF graph per city. This way, the data of each city is interconnected and it is possible to perform complex SPARQL queries to reason over the various dimensions of the data.

---

[7]https://www.elastic.co/products/elasticsearch.

8.5.2   Mobility Mining

This service queries the DM to fetch GPS data about individual private vehicle trajectories. We use a data mining process called *mobility profiling* to extract patterns from these traces. This process takes as input the users' trajectories and computes, in a privacy-preserving fashion, mobility profiles [29] that are then merged via clustering in *routines*. The routines are then mapped to transit network lines, as a sequence of stops, just like other lines of public transport. In this way, we can consider a routing as "alternative bus routes," with a specific schedule, and a probability of existence set to the relative frequency of the routine. These newly introduced routes represent an embedded carpooling service, transparently available in the PETRA application. This virtual bus routes are then stored in the Data Manager to be available to other services.

8.5.3   Data Integration Services

PETRA leverages several data sources, including:

- data provided by transportation agencies, such as GTFS data periodically published;
- canonical map data, such as OpenStreetMap;
- GPS trajectories; and
- historical information, such as user-location information provided by telecom operators.

For each and every data source we created a dedicated service that integrates the original data with a more generic data schema. Note that the system relies on loosely coupled services, which means that each data integration service can be implemented independently and it communicates with the data manager via a well-defined REST API.

The data integration services can be classified according to the use case to which they relate. Specifically, PETRA system supports the following data sources.

The city of Rome, through the public agency Agenzia Mobilità, provides updated open data about its public transport systems. In particular, two main sources of information are offered via its website[8]:

1. Rome public transport GTFS, which is a snapshot of the entire public transport network updated every few weeks and
2. Rome public transport real time API.

---

[8] See http://www.agenziamobilita.roma.it/.

Additionally, Agenzia Mobilità is gathering a large collection of GPS traces from volunteers private cars, used by the mobility mining module, that will be presented in Sect. 8.5.2.

Importing Rome's data relies on an ad-hoc *data acquisition* module (named RDI, Rome Data Importer), that acts as a bridge between the diffcrent kinds of mobility data previously described and thc internal DM.

Given as initial state the Rome public transport GTFS, we can divide the work of the RDI into two sub-tasks: the *daily update* and the *real time update*.

The *daily update* consists of discovering *bus stops routines* and enforcing privacy over them. First the RDI transforms the private car routines into sequences of bus stops and combines them as bus lines: each GPS location is mapped to the closest bus stop within a given radius (in our settings the radius is set to 20 m). Then the RDI further performs data cleaning discarding any bus stop routines consisting of one stop, or only two bus stops which are closer than a given threshold. Note that in order to guarantee car drivers' privacy, the RDI checks if an external attacker could exploit the *bus stops routines* to discover their identity by analysing their vulnerability against the *linking attack* model [24]. To avoid this kind of attack the RDI performs a *privacy risk analysis*. Following the methodology in [3], the result of this method is a probability distribution of the risk of identifying drivers for each routine. If possible, the routines with an identification probability higher than a given *acceptable risk* (refer to [24] for further details) are transformed into a safer version by removing some bus stops, otherwise they are deleted.

Finally all the valid bus stop routines are added to the Rome GTFS data and sent to the DM. Each routine may be used by the JP like any other bus line, even for a portion of the trip. How to make sure the driver of the car can give a ride to the traveler is one of the challenges within the PETRA project. In the *real time update*, the RDI queries the Rome public transport real time API, which consists of a set of XML-RPC methods,[9] which provide updated transport information such as updated (estimated) arrival time, etc., every t minutes, checking for updates (e.g. buses which have been delayed or cancelled) by comparing expected arrival times on the existing GTFS data with real time arrivals. Then it converts possible updates into the GTFS format, and sends them to the DM.

The city of Venice provides similar data. Azienda Veneziana della Mobilità provides through its website[10] an updated GTFS for both bus and steamboats (Venice's main public transport). We created the equivalent of Rome's data imported to daily update Venice's GTFS data into the DM.

---

[9]http://xmlrpc.scripting.com/spec.html.
[10]http://avm.avmspa.it/.

8.5.4   Venice Non-transport Data

In order to provide the activity planning service we required additional data regarding the urban and human components of the city of Venice. To do that, we primarily used the data provided by Telecom Italia for its 2015 Telecom Italia BigData Challenge.[11]

For the city of Venice, the challenge provided telco data of different kinds, plus additional data like a collection of pre-processed tweets, a GIS grid layer over the city of Venice, census data, data coming from Cerved[12] (an Italian information provider). In addition to this list, we used OpenStreetMap,[13] external data about the TIM (Telecom Italia Mobile) market share, and statistics about tourism in Venice from the Municipality of Venice.

In particular, from the BigData Challenge, we used:

- Venice Telecom Grid (GIS shapefile layer), a grid over the city of Venice, defined by Telecom Italia to provide the other data
- "Telecom Demographics" (outbound phone calls per ZIP code, aggregated by cell in the above grid and 15-min time slots)
- "Telecommunication—Calls out", for estimating the crowding level in an area
- "Telecom SocialPulse", a collection of pre-processed tweets from Venice, where POIs and landmarks were replaced with their DBpedia entry—for inferring popularity of POIs
- "Cerved ATECO codes" (selected activity codes for shops, services, etc. in Venice, excluding bars and restaurants), "Cerved Companies" (list of commercial activities corresponding to the above codes) and "Cerved HQ & Branches" (locations of the selected activities), for estimating the list of activities visited along a plan and estimating how many potential customers are served in a time slot.

The BigData Challenge also provided a "Presence" dataset, with an estimated count of people in a cell, in a given time-slot. We decided to use the outgoing calls, nevertheless, as this could be coupled with the demographics dataset to exclude calls performed by people living in the Venice ZIP codes, i.e. non-tourists. This also matches the intuition, confirmed by the municipality of Venice, that Venice residents typically use non-congested pedestrian paths, as their activities differ from the ones performed by the tourists.

As external data, we used in particular:

- Venice OpenStreetMap topography (road-network GIS layer) for computing pedestrian routes
- List of Venice's attractions from OSM for listing attractions grouped by category

---

[11] http://www.telecomitalia.com/tit/en/bigdatachallenge.html.

[12] http://company.cerved.com/?language=en-us.

[13] http://www.openstreetmap.org.

Fig. 8.5 One cell over the city of Venice, with crowding levels ready to be digitalized

- TIM market share
- Statistics on # of daily tourists and useful background information from the Municipality of Venice
- Static pedestrian flow map of Venice, from the Municipality of Venice.

Figure 8.1 shows the entire static map of pedestrial flow intensity in the historical city centre of Venice. In order to use this information, we overlayed the Venice Telecom Grid over it, and split the entire map into different cells. Figure 8.5 shows one of the resulting tiles of the pedestrian map. In each cell, we may see pedestrian paths marked as green (typically empty), orange (intense pedestrian flow), or red (typical place of pedestrian congestion). For each cell we then computed the ratio between the green areas and the total pedestrian area (i.e., green plus orange plus red). This ratio was used to normalize the maximum congestion observed by the outgoing calls from a cell.

8.5.5 Data Analysis

The data analysis module is a set of tools and algorithms used to pre-process the input data in order to extract relevant information, and to extract spatio-temporal patterns of presence in a cell from historical data.

Here we perform the following tasks:

- we overlay the Telecom grid over all the geo-referenced data, like the static pedestrian flow map, and we aggregate street-level information, such as walkable area, etc., up to cell-level;
- we slice time using the same time resolution of the telco data, i.e. 15 min;
- from the static pedestrian flow map, we extract a maximum congestion level which is a function of the red pedestrian areas over the total pedestrian area in a given cell;
- from the telco data, we mine the historical presence in a given cell, in a given time-slot, by aggregating several days of data, after splitting between week-end data and week-day data;
- we pre-process the Social Pulse data by ranking the POIs in Venice by their popularity in the available tweets;

– we pre-process the CERVED data by selecting only relevant business activities
  such as local shops (i.e., excluding restaurants, bars, and services).

The output of this module is then stored into the DM for further uses.

8.6 Case Study: Rome

In the Rome's use case, the PETRA platform, from the traveler's perspective,
provides journey plans from place A to place B. From the operator perspective,
this is done by: importing static and real time urban transport data; merging private
routines into the public transport data; computing uncertainty-aware multi-modal
advices. We here describe the data used in this chapter, how the import step works,
and the results obtained with and without private routines.

The specific application of the PETRA platform for the city of Rome involves a
multi-modal JP that takes into account not only the available transportation systems
in the city (bus, underground, and public bikes) but also privately owned alternatives
like car sharing. This option can be enabled from the JP application that, when
directing drivers to their destinations, inquires if they are willing to collect other
(trusted) users in the vicinity whose destination is the same. The JP application
further stores individual travel preferences, such as total time for each type of
transport and preferred transportation type. When users ask for a route from place A
to B, the returned plan will take into consideration the users' preferences.

8.6.1 Impact of Routines in Journey Planning

We have performed an empirical analysis of our approach, with the objective of
evaluating the impact of adding routines to multi-modal journey planning.

All the information available about the multi-modal transport network is put
together into a knowledge base called a *network snapshot*. We ran the planning
system in two different settings: NoRo, in which the planner uses all the public
transport data available, but no routines; Ro, containing both routines and public
transport data. In each setting, we solved 2000 queries (instances) with the origins
and destinations chosen at random from the logs of the official journey planner
of Agenzia Mobilità. In a query, users can set parameters such as the maximum
walking time per journey m_w, and the maximum number of legs (i.e., segments)
per journey m_l. We set m_w to 20 min, the default planner value. Half of the queries
have m_l set to 5, and the other half is for $m_l = 6$. The public transport data we used
has 8896 stops and 391 routes. Each route is served by a number of trips, to a total
of 39,422 trips per day. The Rome roadmap has 522,529 nodes and 566,400 links.
In the GTFS data, we represent routines with a structure similar to public transport
data. Each routine introduces a new route and a new trip. We started from 1,205,258
GPS trajectories from 262,657 users. After routine extraction, bus mapping, and

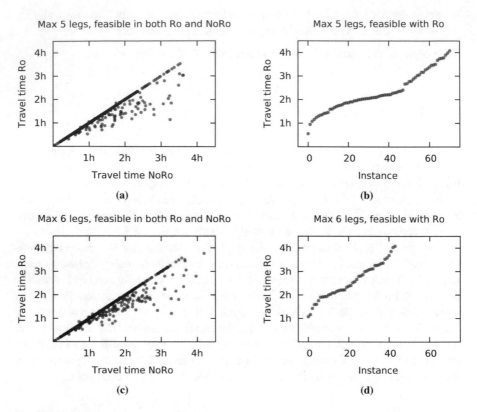

Fig. 8.6 Impact of routines on travel time

anonymization, we ended up with 729 safe mapped routines from 641 users. This increases the number of bus routes to 1120, for a total number of trips of 40,151.

Figure 8.6 illustrates the impact of adding routines as an additional mode. At the left, we compare the travel time in the Ro and NoRo settings. As expected, in a subset of cases, the travel time is the same. On the other hand, all points located below the main diagonal show instances where routines improve the time. In fact, routines can improve both the travel time and the number of legs per journey. The latter has two advantages. First, it makes a trip more convenient to the traveller, as it reduces the number of interchanges. Secondly, it helps increase the set of feasible instances (i.e., instances for which a solution exists). This is important because user-imposed constraints on m_l and m_w can restrict the set of feasible instances. For example, without using routines, in 29.3% of our queries (instances), it is impossible to complete the journey with at most 20 min of walking and at most 5 legs in the journey. Charts at the right in Fig. 8.6 show instances that become feasible after adding routines. When m_l is set to 5, routines are part of the returned plan in 17.5% of the instances. Routines increase the percentage of feasible instances by 7.1%, to a total of 77.8%. In 9.6% of the instances, routines improve the travel time, the average savings per trip being equal to 25.5 min. When $m_l = 6$, routines become part of the plans in 22.3% of the instances. They increase the percentage of feasible instances from 84.5 to 88.9%. In 14.3% of the instances, routines improve the travel time, the average improvement amounting to 22.05 min per trip. Clearly, besides the advantages pointed out, such as the travel time and the number of interchanges in a journey, routines bring additional benefits. These include reducing congestion both on the road and inside public transport vehicles.

8.6.2   Experiments for Planning with Uncertainty

In this section we summarize part of the experiments contained in a short conference paper [8]. We use the same public transport data as in the previous section (i.e., public transport 391 routes, 8896 stops, 39,422 trips per day, 522,529 road map nodes and 566,400 road map segments). No routines are considered in this experiment.

The original data is deterministic. This was extended with a stochastic noise assigned to the original deterministic arrival and departure times. More specifically, for each city we use three distinct network snapshots, one deterministic (i.e., the original snapshot), and two with different levels of stochastic noise. The noise follows a Normal distribution, truncated to a confidence interval of 99.7%. One snapshot has the variance set to $\sigma^2 = 1600$ s, equal roughly to ± 2 min around the original deterministic arrival or departure times. In the other snapshot, we set $\sigma^2 = 6400$ (equal roughly ± 4 min).

We generated 3000 journey plan requests (instances), with 1000 for each of the following departure times: 8 am, 11 am, and 6 pm. The origins and the destinations are picked at random. Trips are constrained to no more than 20 min of walking, and at most 5 legs (segments) per trip. We restrict our attention to 11 am data, a representative subset of the results.

To obtain a deterministic, sequential plan, we have run Dija with a deterministic network snapshot. Uncertainty-aware plans are computed with Dija using a non-deterministic snapshot. Both kinds of plans are simulated in a snapshot with uncertainty, and both the worst-case arrival time and the expected arrival time are measured.

8.6.2.1   Worst-Case Arrival Time

Figure 8.7 compares deterministic and uncertainty-aware plans, in terms of simulated worst-case arrival time. The "main curve" shows the deterministic-plan data.

Depending on how "det" and "non-det" values compare, we distinguish three behaviours in this figure. First off, in a majority of cases, "non-det" points fall on the "main curve", indicating instances where both simulated times coincide. Secondly, in most remaining cases, "non-det" points fall underneath the main

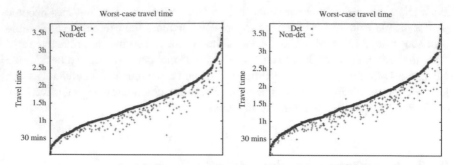

Fig. 8.7 Left: $\sigma^2 = 1600$; Right: $\sigma^2 = 6400$. Sequential plans ("Det") vs contingent plans ("Non-det"). Instances ordered increasingly on the worst-case arrival time of sequential plans

Table 8.1 Contingent vs deterministic plans: key statistics with worst-case travel time

DP	ASM	ASP	DP	ASM	ASP
$\sigma^2 = 1600$			$\sigma^2 = 6400$		
23.85%	16.71	18.08%	32.70%	18.10	18.35%

DP = percentage of cases when differences occur. ASM = average savings per trip, in minutes, when differences occur. ASP = average savings per trip, as a percentage of the trip time, when differences occur

curve, corresponding to cases where contingent plans have a better simulated time. Table 8.1, discussed later, shows exact percentages and other key statistics corresponding to this behaviour. Thirdly, in just a few cases, deterministic plans have a better arrival time. Part of the explanation is that the Dija planner optimizes plans on a linear combination of the number of journey legs (segments) and the arrival time. In a few cases, the optimal contingent plan has a longer arrival time, and fewer legs than the corresponding deterministic plan. See details later in this section about how the weights of the linear combination are chosen.

We conclude from Fig. 8.7 is that uncertainty-aware plans can often help arrive at the destination earlier.

Table 8.1 shows key statistics of the comparison. Header DP shows the percentage of "Non-det" dots not placed on the main curve in Fig. 8.7. As expected, increasing the level of uncertainty increases the DP value. Remarkably, ASM and ASP remain stable when σ^2 varies.

8.6.2.2 Expected Arrival Time

On this metric, differences between contingent and deterministic plans are smaller, and they go in both directions. This is consistent with the planner's optimization strategy: the worst-case cost is the main criterion, and the expected cost is for tie-breaking.

In Rome, when $\sigma^2 = 1600$, in 7.03% of the cases we see an advantage for contingent plans, whereas 7.49% of all cases favour deterministic plans. When $\sigma^2 = 6400$, the numbers change to 12.63% and 8.53%. The other times of the day considered (8 am and 6 pm) show similar results.

8.6.2.3   Dynamic Deterministic Replanning (DDR)

We have implemented a strategy that performs, in every state, a deterministic replanning, but simulates the first leg of each plan under a snapshot with uncertainty. We have measured the simulated worst-case arrival time. DDR is better than deterministic planning, but not as good as contingent planning. For example, when comparing DDR to contingent plans, the percentage of cases favourable to contingent plans is 9.20% when $\sigma^2 = 1600$ and 12.44% when $\sigma^2 = 6400$. The percentage of cases favourable to DDR is smaller, being 0.70% when $\sigma^2 = 1600$ and 1.53% when $\sigma^2 = 6400$.

8.6.2.4   Search Time

As expected, the search time increases with σ^2. A max limit of 30,000 expanded nodes was set in experiments. With the most difficult uncertainty level ($\sigma^2 = 6400$), Rome instances are solved in 98% cases. In successful cases, the average search times in seconds, measured on a 2.7 GHz Ubuntu machine, are: 0.01 ($\sigma^2 = 0$), 0.08 ($\sigma^2 = 1600$) and 0.18 ($\sigma^2 = 6400$). A more detailed discussion is beyond the scope of this chapter.

8.6.2.5   Weights in the Linear Combination of a Plan Cost Formula

The cost of a branch in a journey plan is a linear combination of two factors: the travel time t, measured in seconds, and the number of legs l along that branch:

$$C = cw \times t + (1 - cw) \times l.$$

Figure 8.8 illustrates the impact of the weight cw on the CPU time (top) and the worst-case travel time (bottom). At the left, there is little difference between using $cw = 1$ and $cw = 0.1$, in terms of CPU time and worst-case travel time. In the middle column of the figure, using $cw = 0.005$ reduces the CPU time in the case of many difficult instances, as compared to the benchmark setting $cw = 1$. There is little difference in terms of worst-case travel time between $cw = 0.005$ and $cw = 1$. Even when worst-case travel time increases slightly, the number of legs in the journey can actually decrease. In the rightmost column of the figure, observe that

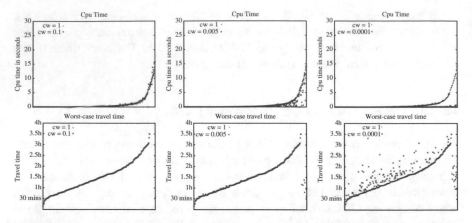

Fig. 8.8 A few sample values for cw, compared against a benchmark setting $cw = 1$, which optimizes plans purely on the travel time

using a value such as $cw = 0.0001$ decreases the CPU time even further. However, the worst-case travel time can also increase. Based on such an analysis, the default value for cw is set to 0.005.

8.7 Case Study: Venice

We present here the use case in the city of Venice, and the key aspects of crowd balancing and journey planning, which constitute the problem solved by togetThere.

8.7.1 Crowd Balancing

Consider that the map of a city is partitioned into a set of smaller components. For instance, components could result from a grid discretization of the city area, with cells having a fixed or a variable size. In a finer-grain decomposition, the components could be smaller entities such as street segments.

For each component, define a maximum congestion level. Also, at any given point in time, the component at hand has an actual congestion level. As an actual congestion level approaches the maximum congestion level, actions such as walking through the component at hand is negatively impacted. For example, on the narrow streets of Venice, a high congestion level can significantly reduce the speed of walking, or even make it impossible to traverse a given area. For a given component, we compute the walking speed as a function to its actual congestion level and max congestion level.

In computing the walking speed v across a component, we adopt Bruno and Venuti's approach [13], using the following speed–density formula:

$$v = v_c \left(1 - e^{-\gamma\left(\frac{1}{u} - \frac{1}{u_c}\right)}\right),$$

where v_c represents the walking speed under no congestion, u is the current congestion level, and u_c is the maximum congestion level. See the empirical evaluation in Sect. 8.7.4 for a range of values used for the γ parameter.

In addition, we assume that points of interest can have a maximum and an actual congestion level. We compute the time t to visit a point of interest in a similar manner, as a function of the current congestion level and the maximum congestion level at that point of interest:

$$t = t_c \left(1 - e^{-\gamma\left(\frac{1}{u} - \frac{1}{u_c}\right)}\right).$$

In the previous equation, t_c is the visiting time under no congestion, u is the actual congestion level, and u_c is the maximum congestion level.

Crowd balancing aims at distributing a group of individuals across the city so that actual congestion levels do not exceed a threshold that would impact the travel speed or the visiting time at a point of interest.

8.7.2 The Planning Problem

The role of planning is to produce activity plans for individual tourists. Informally, an activity plan has to specify what points of interest to visit, when to visit them, and how to travel from one point of interest to the next.

As such, an activity plan is an ordered sequence of locations to visit, together with a travel plan between any two consecutive such locations. The locations are a starting point, such as the hotel of the user, an ordered set of points of interest to visit, and a final destination at the end of the plan. Notice that points of interest are not limited to tourist objectives, including, for instance, restaurants and shops as well. Often, the start and the final destination coincide, as it is the case when the user returns to the hotel at the end of the day. We distinguish between travel actions and visit actions in a plan. Examples of "visit" actions include enjoying a tourist attraction, stopping to a restaurant for lunch or performing shopping in a store. Actions have temporal information that specifies the times to start and complete the action at hand.

Individual activity plans take into account the impact of previously computed plans onto the crowding levels across various areas of the map. As such, a new individual activity plan will avoid going into overcrowded areas. This allows to achieve a crowd balancing effect when putting all individual plans together.

The input to the planning engine includes a set of points of interest selected by the user. Each point of interest can be characterized by constraints such as an opening-hours interval and the amount of time needed to visit the point of interest at hand. In addition, the input contains information about the transport network, such as roadmap information, and the routes and schedules of public transport services. We call all the transport-network data available a snapshot network.

An activity plan is valid if it includes all the points of interest given as input, the constraints associated with the points of interest are satisfied, and the travel leg between each pair of two consecutive points of interest is a valid journey plan. A journey plan is valid if all its steps can be performed according to the information encoded in the transportation network snapshot given as input.

The plan-quality criterion considered in this work is the time needed to complete the activity plan. Thus, the congestion levels of various areas across the map directly impact the quality of a plan. A planning engine that aims at minimizing the plan-execution time will prefer travel legs with a faster travel time, which in turn may result in avoiding congested, low-speed areas.

8.7.3 Journey and Activity Planning

In this section we describe our approach to computing the activity plans. Recall that an activity plan contains an ordered sequence of locations to visit (start, POIs, destination), the journey plans connecting each pair of two consecutive locations, and temporal annotations.

In designing our planning system, we decided to trade away the optimality for a stronger scalability. As such, as shown in Algorithm 1, the planning problem is decomposed into two sub-problems, namely finding a good ordering of the points of interest and finding a detailed journey plan between each pair of consecutive points of interest in the ordering.

Algorithm 1 Planning approach

Require: Origin s, POI list L, transport network snapshot N, travel-time matrix m
1: $O \leftarrow$ computeCandidateOrderings(s, L, m)
2: $\pi \leftarrow$ computePlan(s, O, N, m)
3: **return** π

8.7.3.1 Computing Candidate Orderings

The role of the method computeCandidateOrderings is to select a subset of orderings for the POIs contained in the list L. This is necessary because, as shown later in this section, computing a plan involves a search in a graph whose size depends on the number of candidate orderings considered.

When the number of POIs to visit in one day is small (e.g., 5), even a brute-force enumeration of possible orderings would be feasible. However, in many realistic scenarios, the number of locations to visit n can be significantly higher. When $|L|$ is large, a search space considering all $|L|!$ possible orderings is impractical. Pruning can be employed to reduce the number of candidate orderings to a subset.

For simplicity, in this section we assume that the two endpoints of the plan trajectory (start and destination) coincide. Thus, finding a good ordering of the points of interest is equivalent to finding a solution to a TSP problem where the nodes to consider include the starting point and the points of interest to visit. An ordering begins with the starting location, continues with all POIs, and ends with the starting location as well.

In selecting a subset of orderings, we define a TSP instance, solve it with a suboptimal but scalable tabu search [15], and return the top K solutions, where K is a parameter. The choice of restricting our attention to only the top K orderings returned by the tabu search greatly reduces the number of possible combinations, while generally preserving the quality of the results. To define the TSP instance, besides the set of locations (start/destination and POIs to visit) we need a matrix with the travel times between any pair of nodes.

We use a static, pre-computed matrix m with estimated travel times between any pair of nodes in the TSP problem. For walking legs, these travel times take into account the distance on the road map, but they ignore at this stage any congestion level. We found this to be a good tradeoff, as a static pre-computed matrix allows to quickly look up the travel times needed. Note that using estimated travel times at this stage of plan computation may affect the quality of the computed orderings, but it does *not* affect the correctness of the plans provided in the end. The reason is that the only role of this routine is to output one or several candidate orderings. Then, in a subsequent part of the plan computation process, described later in this section, plans based on various orderings will be computed using the most accurate travel time information available.

Tabu-search is typically known due its memory structure called tabu-list that avoids to get back to already visited solutions in the last iterations (the diversity of the explored solutions increases as compared to a local search algorithm). Based on the number of the selected POIs, we set the tabu-list size to $|L| - 1$, and the number of total iterations to $4 \times (|L|-1)^2$. The initial TSP solution considers all the selected POIs randomly ordered. Only the start and the end positions are constrained to the explicit selections made by the user. At each iteration, the tabu search moves to the best neighbouring solution, even if it is worse than the current one, it stores it into the tabu list, and avoids all the neighbouring solutions contained in the tabu-list. The number of the solutions to return K is set to 10 in our experiments.

8.7.3.2 Computing Plans for a Given Set of POI Orderings

The search space explored to find plans is a directed graph. Each node is a partial sequence of locations (start and POIs), originating at the starting point. In a transition (directed edge), the successor node has a sequence of locations similar to the parent's sequence, except that the successor's sequence is one step longer. The subset of POI orderings control the size of the search space.

More precisely, let $N = \{s\} \cup L$ be the set of locations to consider (i.e., the start together with the POIs), and let O be a set of orderings of these locations. Given an ordering $o \in O$, we call the l-prefix of o the sub-sequence $o_{|l}$ of o restricted to the first l elements. For every $o \in O$ and for every $l \leq |N|$, the l-prefix $o_{|l}$ defines a node in the search space. When two distinct orderings o and o' have an identical l-prefix, only one node is defined in the search space. Two nodes n and n' are connected with a directed edge (transition) from n to n' iff $\exists o \in O, \exists l < |N|$ such that n corresponds to $o_{|l}$ and n' corresponds to $o_{|l+1}$.

The cost of a transition is the travel time from the second last location to the last location in the sequence of locations at hand. This cost is computed dynamically, using the most accurate data available about the status of the transportation network. In particular, walking times take into account the known congestion levels. Besides transitions, POIs have a time cost associated with visiting the POI at hand. Similarly to travel times, these visiting times are computed dynamically, taking into account the congestion levels at the time of the visit.

Note the impact of using a subset of orderings onto the size of the search space. The full search space has a number of nodes in the order of $|L| \times |L|!$ nodes, whereas a restriction to K orderings results in a search space with a number of nodes in the order of $|L| \times K$.

The search space is explored with the A* algorithm [17], starting from the root node whose sequence of locations has only the start location. Any node whose sequence contains a complete cycle (i.e., begins with the start, continues with all POIs, and ends with the start) is a goal state. The A* algorithm uses a heuristic function h that guides the search. Given a node n, $h(n)$ estimates the cost from the current node n to any goal node. When $h(n)$ never over-estimates the smallest possible cost to reach a goal from n, h is said to be admissible. Using an admissible function ensures that A* computes optimal solutions in the search graph given as input. In our application, an easy to compute admissible heuristic is $h(n) = \arg\max_p m[p, s]$, where m is a pre-computed travel-time distance with the property that $m[i, j]$ never over-estimates the true travel time from i to j, and p is iterated over all POIs not included in the ordered sequence of locations corresponding to node n.

Given the departure location and time of the tourist and her expected arrival location, the TSP problem among all the selected POIs is solved by taking into account the traveling time affected by the traveling distance only. The set of the first *K-TSP* solutions (the K-shortest/K-fastest paths) is stored for further computations.

The directed graph of the K-TSP solutions,[14] with the departure location as the root node and the arrival location as the target node, is built. The weights on the edges represent the dynamic traveling time between any pair of POIs, which is computed by taking into account the crowding level on the walking path that connects them and the occupancies at the arrival's POI. Expected arrival times at POIs and their expected visiting times (that are affected by the POI's occupancy) are crucial to the selection of the right, time-based, crowding level.

The graph is traversed with an instance of the A* algorithm, which builds the fastest sequence of activities. As the speed of travel and the duration of visiting a POI directly depends on the congestion level along the travel legs and at the POIs, plans with a shorter duration may correspond to taking less crowded routes and visiting less crowded POIs.

8.7.3.3 Dynamic Replanning

The recommended visit itinerary is provided to the tourist, which can then begin her journey. At the end of the visit to each POI included in the current plan, a re-planning request is sent to our planning system. In these cases, the planning system computes a new plan, as shown earlier, but with the input data adapted to the new state of the user. Specifically, now the starting location is the current location of the user (i.e., the POI at hand), and the locations to visit are all POIs selected to visit in the initial planning round (e.g., at the beginning of the day), except for those visited so far.

The re-planning allows to update the recommendation if expected crowding levels or occupancies during the day have changed due to the requests received since the last computation performed for the same tourist. Furthermore, crowding levels and occupancies are updated following the changes between the newly computed itinerary and the previous one.

8.7.4 Experiments

8.7.4.1 Simulation Setup

In order to simulate tourism activity in a realistic way, we used background information from the Mobility Agency of Venice in conjunction with the data coming from the BigData Challenge. In particular, we learned from the Mobility Agency that Venice expects a minimum of 60,000 tourists in a day, up to a peak

---

[14]The choice of solving the *K-TSP* problem was driven by the need of reducing the complexity of solving the A* algorithm. This approach lets to prune the full graph connecting the full set of selected POIs, which would have $O(N!)$ complexity (where N is the amount of selected POIs), to a graph with traversing complexity of $O(N^K)$.

of 250,000. Thus we generated demand for 60,000 users of our system. For each demand, we had to include a set of points of attraction. Based on the information in our possession, we picked the number of points of interest to include in a visit sequence from a normal distribution with the mean equal to 5. The specific set of POIs to include was chosen from the list of POIs found in the SocialPulse dataset, picking a specific POI with a probability proportional to the frequency of a given POI in the dataset. The duration of the visit at a specific POI or landmark was estimated using the POI type (i.e., museum, church, monument, etc.), and the background information in our possession.

We then generated 60,000 visit requests, and solved each of them with toget-There, as well as with two baselines: a greedy shortest path algorithm, and a greedy approach [11] typically followed by tourists, i.e. going to the "best attraction" first, then proceeding to the remaining ones. Each solution returned by the system consisted in a time-stamped walking path, with time of arrival and departure from each location.

For γ, we chose the value of 0.245 as reported in [13], plus we made it vary from 0.2 to 0.3 with an increment of 0.01, to enlarge the search space.

As baselines approaches to comparing a travel segment, we chose a method based on the shortest path, and one based on a greedy approach reaching the closest location(s) first [11]. In the shortest path approach, each tourist is routed independently from the others, following just the shortest path among all the POIs to visit. In togetThere, each tourist is routed according to the forecast crowding level, and according to the projection of all the previously started plans.

8.7.4.2 Simulation Results

Figure 8.9 shows the comparisons between the returned visit plans by togetThere vs. shortest path (in (a)) and by togetThere vs greedy (in (b)). Both the plots were generated with $\gamma = 0.245$ in togetThere. Each dot represents a user request. Its coordinates represent the ratio of the total durations (x axis) and the ratio of the total distances (y axis). More formally, let $tcrowd$ be the total duration of the plan returned by togetThere, $lcrowd$ its total distance, $tshort$ the total duration of the plan returned by the shortest path based approach, $lshort$ its total distance, $tgreedy$ be the total duration of the plan returned by the greedy based approach, $lgreedy$ its total distance, then the x axis in (a) is computed as

$$\left(\frac{tcrowd}{tshort} - 1 \right) \times 100$$

the y axis in (a) is computed as

$$\left(\frac{lcrowd}{lshort} - 1 \right) \times 100$$

Fig. 8.9 Comparing output
of togetThere with shortest
path (**a**) and togetThere with
greedy (**b**). In both plots, γ is
set to 0.245 in togetThere.
Each dot is a visit request: the
x axis is the ratio of total
duration; the y axis is the
ratio of total distance

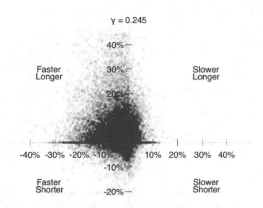

(a) togetThere vs shortest path

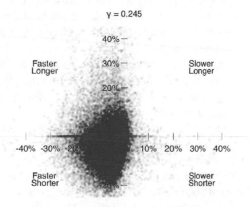

(b) togetThere vs greedy

the x axis in (b) is computed as

$$\left(\frac{tcrowd}{tgreedy} - 1\right) \times 100$$

and finally the y axis in (b) is computed as $\left(\frac{lcrowd}{lgreedy} - 1\right) \times 100$. As we can see, the
majority of the dots are in the left half of both the plots, with a larger proportion in
(b). In (a), the majority of the dots is expected to be in the upper left quadrant, i.e.
the plans returned by togetThere should be faster (this is obtained by design) but
longer (this is the price to pay to have faster plans than the shortest ones). We do
have a small number of dots in the bottom half, which may seem counter-intuitive
as we cannot beat the shortest path in length, by design. However, recall that we are
not executing a pure shortest path approach as a baseline, but rather a greedy version

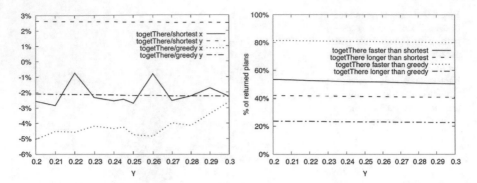

Fig. 8.10 Percentage of either faster or longer plans returned by togetThere vs the baselines. (**a**) Ratios for all values of γ. (**b**) % of faster or longer plans

of it, considering only the top k shortest paths. This does not hold in (b), where we do have room to be not only faster, but also shorter than the greedy approach based on proceeding always towards the closest next point of interest.

We had similar results for different values of γ. In order to better appreciate these results, for each scatter plot of this kind, we computed the centroid. The desired result should have the x coordinate less than 0, while the y coordinate is expected to be greater than 1 when comparing with the shortest path, and less than 1 when comparing with the greedy approach.

Figure 8.10a shows the trend of centroid coordinates by varying γ. As we see, the y coordinate (i.e. the ratio of total distance) stays quite constant, while the x coordinate fluctuates a bit more, with a slightly increasing trend. Despite this, we report no major impact of varying γ over the average gain in total duration. Figure 8.10b shows instead the percentage of plans returned by togetThere that are either faster than the ones returned by the shortest path or longer than them, with the same computations repeated also for the comparison with the greedy approach. In this plot, the almost constant trend of gain by using togetThere instead of one of the baselines is even more evident. We also note that, with these values of γ, we have the majority of plans faster than the shortest ones.

Given that the plans returned by togetThere are faster, how much additional time do tourists typically have? We report that an almost constant (by varying γ) 20% of plans will complete at least 15 min earlier than their counterpart provided by the shortest path. Roughly 14% of the plans will be at least 20 min faster, while about 7% of the plans are at least half an hour faster. This is enough time for one additional break during the day, or a brief visit to one additional landmark, or, as we see in the next section, to do more shopping.

8.7.4.3   Business Impact

Besides leaving the tourists with a better experience, and helping the city reduce the cost and inconveniences caused by pedestrian congestion, the city of Venice could see some unexpected additional benefits. First, let us summarize the features of the majority of plans returned by togetThere, when comparing them with either the shortest paths or the greedy ones:

- our plans are faster: this leaves more free time;
- our plans are longer than the shortest path: this means tourists see more landmarks, or business activities, during their visit;
- our plans reduce congestion, also by moving people away from over-crowded areas.

Can we exploit this with a return on the local economy, without adding any more tourists to Venice? We performed an experiment aimed at assessing the impact on local business activities of our smarter plans. To better understand the rationale behind this experiment, we need some background on the typical business activities in the city centre in Venice. If we exclude restaurants, bars, and other services, and take into account only shops, we may realize that Venice is full of small-size, typically family-ran shops selling handicrafts like glass items, jewelry, fabric items, and so on. These shops typically extend to 30–40 m^2, with one or two persons attending to customers, with sudden crowds entering the shop, causing the owners to either lower the quality of service per customer, or temporarily refuse customers when reaching a certain thresholds of customers per time unit.

We designed our experiments as follows: first, according to the background knowledge in our possession, we estimated that a typical shop may serve with a good quality of service up to 15 people in a 15 min time slot (we chose to divide time according to the same time slots we had in the rest of the data). This also matches the estimate reported in [30]. Any customer entering the shop over this threshold falls within the "saturated" shopping, which the shop cannot serve within that time slot, with an acceptable quality of service. Then, thanks to the availability of the CERVED dataset of business activities in Venice, we mapped each of the returned plans to all the shops within 15 m from the walking path, thus estimating which shops each tourist may potentially visit. In other words, we computed the number of "impressions" received by each shop in a 15 min time slot. We then looked at the saturated shops in the shortest path plans, and compared with the plans returned by togetThere. Given the features as summarized above, we report a significant decrease in the number of saturated shops, as, by moving people to less crowded areas, we basically move people from above the saturated threshold, to below that, by having them see (and potentially visit) different business activities. Moreover, thanks to the longer plans, we actually also increase the number of shops visited in a plan. Lastly, by giving more free time in the end, we could increase the local economy even more, if the additional free time is spent for shopping.

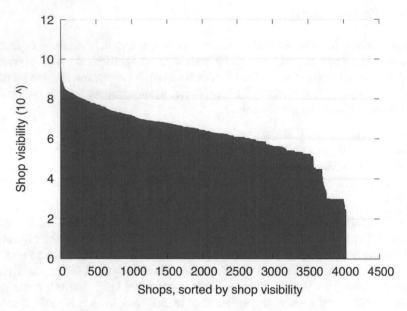

Fig. 8.11 Shop visibility

Without considering the additional free time that could potentially be spent for shopping, we have then counted the impressions for 4035 business activities in the city centre of Venice, reported in Fig. 8.11. Using the crowd steering strategy the shop visibility increases by 4.5%.

8.8   Conclusion

We have introduced an integrated platform to supply urban travelers with smart journey and activity advices, on a multi-modal network. We have presented the architecture of the platform and its main components. We have focused on two use cases, one on Rome and another in Venice.

In Rome, we have evaluated multi-modal journey planning when variations in the arrival and departure times of buses could impact a multi-leg trip. In addition, we have demonstrated the benefits of adding routines (stemming from private car trips available for car sharing) to a multi-modal transportation network.

In Venice, we have focused on tourist activity and journey planning, where a tourist visits multiple points of interests during a day. We have shown the benefits of a planning approach aiming at balancing the crowd levels in a crowded city such as Venice.

In future work, we plan to take into account additional sources of uncertainty in a multi-modal transport network. Introducing additional transport modes, such as cars available for hire in a shared network, or electrical vehicles, would be another interesting direction.

Acknowledgements This work has been partially supported by the EC under the FET-Open Project n. FP7-ICT-609042, PETRA.

References

1. R. Baraglia, C.I. Muntean, F.M. Nardini, F. Silvestri, Learnext: learning to predict tourists movements, in *CIKM '13* (ACM, New York, 2013), pp. 751–756
2. H. Bast, D. Delling, A.V. Goldberg, M. Müller-Hannemann, T. Pajor, P. Sanders, D. Wagner, R.F. Werneck, Route planning in transportation networks. CoRR, abs/1504.05140 (2015)
3. A. Basu, A. Monreale, J.C. Corena, F. Giannotti, D. Pedreschi, S. Kiyomoto, Y. Miyake, T. Yanagihara, R. Trasarti, A privacy risk model for trajectory data, in *Trust Management VIII* (Springer, Berlin, 2014)
4. M. Berlingerio, F. Calabrese, G. Di Lorenzo, R. Nair, F. Pinelli, M.L. Sbodio, AllAboard: a system for exploring urban mobility and optimizing public transport using cellphone data, in *ECML PKDD 2013* (Springer, Berlin, 2013), pp. 663–666
5. M. Berlingerio, F. Calabrese, G. Di Lorenzo, X. Dong, Y. Gkoufas, D. Mavroeidis. Safercity: a system for detecting and analyzing incidents from social media, in *(Demo paper) IEEE '13 Workshops* (2013), pp. 1077–1080
6. M. Berlingerio, F. Calabrese, G. Di Lorenzo, R. Nair, F. Pinelli, M.L. Sbodio, Allaboard: a system for exploring urban mobility and optimizing public transport using cellphone data, in *(Demo paper) ECML-PKDD '13* (2013), pp. 663–666
7. M. Berlingerio, B. Ghaddar, R. Guidotti, A. Pascale, A. Sassi, The graal of carpooling: green and social optimization from crowd-sourced data. Trans. Res. Part C: Emerg. Technol. **80**, 20–36 (2017)
8. A. Botea, S. Braghin, Contingent versus deterministic plans in multi-modal journey planning, in *Proceedings of the 25th International Conference on Automated Planning and Scheduling (ICAPS)* (2015), pp. 268–272
9. A. Botea, E. Nikolova, M. Berlingerio, Multi-modal journey planning in the presence of uncertainty, in *ICAPS* (2013)
10. E. Bouillet, L. Gasparini, O. Verscheure, Towards a real time public transport awareness system: case study in Dublin, in *ACM Multimedia*, ed. by K.S. Candan, S. Panchanathan, B. Prabhakaran, H. Sundaram, W.C. Feng, N. Sebe (ACM, New York, 2011), pp. 797–798
11. I.R. Brilhante, J.A. Fernandes de Macêdo, F.M. Nardini, R. Perego, C. Renso, Where shall we go today?: planning touristic tours with tripbuilder, in *CIKM '13* (ACM, New York, 2013), pp. 757–762
12. I.R. Brilhante, José A.F. de Macêdo, F.M. Nardini, R. Perego, C. Renso, Tripbuilder: a tool for recommending sightseeing tours, in *ECIR 2014* (2014), pp. 771–774
13. L. Bruno, F. Venuti, The pedestrian speed-density relation: modelling and application, in *Proceedings of the International Conference Footbridge* (2008)
14. P. Ge, W.M. Zheng, X.H. Xiao, Y.Q. Qiu, P.Y. Ren, Simulation study based on the regional space-time load balancing of jiuzhaigou. J. Ind. Eng. Eng. Manag. **27**(2), 115–124 (2013)
15. F. Glover, Tabu search – part I. ORSA J. Comput. **1**(3), 190–206 (1989)
16. R. Guidotti, A. Sassi, M. Berlingerio, A. Pascale, B. Ghaddar, Social or green? A data-driven approach for more enjoyable carpooling, in *IEEE 18th International Conference on Intelligent Transportation Systems, ITSC 2015*, Gran Canaria, 15–18 September 2015, pp. 842–847

17. P. Hart, N. Nilsson, B. Raphael, A formal basis for the heuristic determination of minimum cost paths. IEEE Trans. Syst. Sci. Cybernet. **4**, 100–107 (1968)
18. H.-P. Hsieh, C.-T. Li, Mining and planning time-aware routes from check-in data, in *CIKM '14* (ACM, New York, 2014), pp. 481–490
19. M. Jin, D. Zheng, P. Ren, Application of RFID based low-carbon scenic integrated management system in Jiuzhaigou area, in *Proceedings of the Seventh International Conference on Management Science and Engineering Management* ed. by J. Xu, J.A. Fry, B. Lev, A. Hajiyev. Lecture Notes in Electrical Engineering, vol. 242 (Springer, Berlin, 2014), pp. 1269–1280
20. L. Jost, Entropy and diversity. Oikos **113**(2), 363–375 (2006)
21. A.M.W. Leong, X. Li, A study on tourist management in china based on radio frequency identification (rfid) technology, in *Digital Culture and E-Tourism: Technologies, Applications and Management Approaches: Technologies, Applications and Management Approaches* (2010), p. 190
22. S. Lin, Y. Yuan, W. Zheng, P. Ren, M. Jin, Evaluation model construction and simulation research on tourist diversion strategy in ecotourism scenery sport. Appl. Math. Inform. Sci. **7**(6), 2249–2257 (2013)
23. M. Mamei, F. Zambonelli, Field-based approaches to adaptive motion coordination in pervasive computing scenarios, in *Handbook of Algorithms for Mobile and Wireless Networking and Computing* (CRC Press, Boca Raton, 2004)
24. A. Monreale, G.L. Andrienko, N.V. Andrienko, F. Giannotti, D. Pedreschi, S. Rinzivillo, S. Wrobel, Movement data anonymity through generalization. Trans. Data Priv. **3**(2), 91–121 (2010)
25. M.A. Peot, D.E. Smith, Conditional nonlinear planning, in *Proceedings of the First International Conference on Artificial Intelligence Planning Systems* (Morgan Kaufmann Publishers Inc., San Francisco, 1992), pp. 189–197
26. A. Popescu, G. Grefenstette, P.-A. Moëllic, Mining tourist information from user-supplied collections, in *CIKM '09* , November 2009 (ACM Press, New York, 2009), pp. 189–197
27. Y.Q. Qiu, P. Ge, P.Y. Ren, A study on temporal and spatial navigation based on the load-balance of tourists in jiuzhaigou valley. Resour. Sci. **32**(2), 118–123 (2010)
28. C. Song, Z. Qu, N. Blumm, A.-L. Barabsi, Limits of predictability in human mobility. Science **327**(5968), 1018–1021 (2010)
29. R. Trasarti, F. Pinelli, M. Nanni, F. Giannotti, Mining mobility user profiles for car pooling, in *KDD '11* (2011)
30. VV. AA, Effective crowd management: guidelines on maintaining the safety and security of your customers, employees and store. Report, NRF National Retail Federation (2005), http://NRF.com/crowdmanagement
31. W. Zheng, P. Ge, M. Jin, P. Ren, H. Gao, The simulation study of space-time shunt in jiuzhaigou based on bottleneck theory. Appl. Math. Inf. Sci. **6**(3), 993–1000 (2012)

Chapter 9
Mobility Pattern Identification Based on Mobile Phone Data

Chao Yang, Yuliang Zhang, Satish V. Ukkusuri, and Rongrong Zhu

9.1 Introduction

Understanding human mobility pattern is a crucial component of urban planning and has applications in analyzing the dynamics of cities, land use changes, and epidemic control. With economic growth and rapid advances in sensing technology, mobile phone ownership and usage is increasing. In China, the number of mobile phone users is close to 1.35 billion by April 2017. Many researches realized that the mobile phone data can be used as an important complement of the existing traffic data collection technology [1–3] in human mobility study. In trip origin–destination (OD) matrix generation, White and Wells [4] obtained the OD matrix with the MOLA data (OD matrix which was obtained from a roadside survey conducted in 1992) and phone calls cost data. Combining mobile phone signaling data with vehicle detection data, Friedrich et al. [5] obtained vehicles OD matrix by identifying the vehicle on the roads using fuzzy algorithm and generating the travel path using Kalman filter. This method can be used for continuous monitoring of road service and network traffic demand. Recently, many researchers focus on the mobility pattern and activity model of human. For frequency trajectory identification, there is significant work on the pattern mining using GPS data, and development of indices such as distance, slope, and spatial similarity to measure the

C. Yang (✉) · Y. Zhang · R. Zhu
School of Transportation Engineering, Key Laboratory of Road and Traffic Engineering of the Ministry of Education, Tongji University, Shanghai, China
e-mail: tongjiyc@tongji.edu.cn

S. V. Ukkusuri
School of Transportation Engineering, Key Laboratory of Road and Traffic Engineering of the Ministry of Education, Tongji University, Shanghai, China

School of Civil Engineering, Purdue University, West Lafayette, IN, USA

© Springer International Publishing AG, part of Springer Nature 2019   217
S. V. Ukkusuri, C. Yang (eds.), *Transportation Analytics in the Era of Big Data*,
Complex Networks and Dynamic Systems 4,
https://doi.org/10.1007/978-3-319-75862-6_9

frequency patterns [6–9]. But, they only consider the spatial–temporal trajectory and have not considered the meaning of location for the users. Song et al. [10] calculate the chaotic degree of personal mobile trajectory (entropy) by the anonymous mobile phone users and find that 93% of the users are predictable. This research gives us confidence that it is possible to predict the users' future travel using historical data. Ahas et al. [11] propose a way to use passive mobile phone data (data generated by call and message) to define meaningful points (home point, work point, and secondary point) for users. Phithakkitnukoon et al. [12] use "activity-aware map" to estimate the activities most likely related to the specific space and then build a simple model to describe the activity type of the users. Hasan and Ukkusuri [13] use check-in data to classify the urban activity pattern using topic models. Kung et al. [14] identify the home/work location and analyze the commute mobility using the mobile phone record data and compare the results of different cities. Farrahi and Gatica-Perez [15] use latent Dirichlet allocation (LDA) model to discover the location (home, work, and other) routines of the 97 mobile phone users. They build location sequence bag to represent the mobility information of days, and use topics to explain the mobility patterns. But, 200 topics of their model are too many to explain and it is hard to model the mobility by single topic because the mobility pattern of the day is represented by the distribution of all the topics even though some days only show one topic.

In this research, we first identify home and work locations for the mobile phone users. Following the work of Farrahi and Gatica-Perez [15] and Shih et al. [9], we build the bag of location sequence for all days of users. Then, we develop an LDA model to analyze the location sequence information of the users. We cluster the model results to decipher the mobility patterns of the users and compare the different mobility patterns of the users on weekday and weekend. Finally, representative daily location sequence is captured for each pattern and by measuring the accuracy of the representative feature, we find that the representative mobility feature of cluster can describe the main mobility of the users to a big degree.

9.2 Data and Methodologies

In this study, we use 60 days of the call record data (CRD) of Shenzhen city, China in August, September, and October in 2013. Data of few days were missing. The base station regions (BSRs) defined by the Voronoi diagram are illustrated in Fig. 9.1, and BSRs are used to locate users [16]. Positioning accuracy ranges from 100 m to 2000 m depending on the density of the base station. There are totally 3884 BSRs in Shenzhen city. Samples of the CRD are listed in Table 9.1. Data cleaning is conducted due to lack of field information, matching error with BSRs, wrong IMSI, and duplicate records.

Fig. 9.1 Base station region (BSR) by Voronoi. Fine line is the boundary of the BSR and coarse line is the boundary of the district region in Shenzhen

Table 9.1 Sample of the mobile phone call record data (CRD)

IMSI	BSC	Cellular ID	Sector ID	Call sign	Data and time
4600357544****4	13	200	2	0	2013/08/19 05:45:58
4600357544****1	18	1009	2	0	2013/08/19 23:58:04
4600357544****0	14	131	1	0	2013/08/19 18:50:34

IMSI is the unique sim card ID of user, BSC refers to base station controller. With BSC, cellular ID, and sector ID, the BSR of user location can be identified. For call sign, 0 means dialing, 1 means incoming call, 2 means hard handover, and 3 means null value

9.2.1 Identification of Home/Work Locations

To identify the home and work location, we only choose the data on weekday (41 days). It is considered that the span of 2 weeks (for weekday, 10 days) can relatively show user's mobility pattern rule well in the general case, so we choose the users who both have call records during home time (8 pm–8 am) and working time (9 am–6 pm) for more than 10 weekdays.

Identification of the home/work locations is based on our daily behavior habits. Residents' activity starts from home and ends at home. During the daytime, residents (for commuters) are more likely to stay in their work locations, so most of the calls in working time are made at work locations. The rules of identifying one user's home and work locations are presented below (some definitions are showed in Fig. 9.2):

- For home location:

  - Days are demarcated by 3:00.
  - Record the earliest call (after 3:00) and the latest call (before 3:00) of each day.

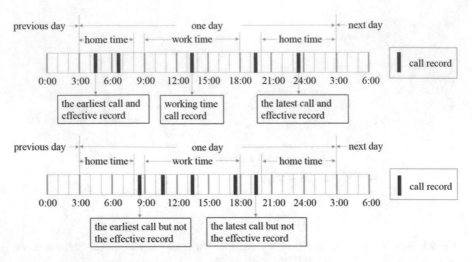

Fig. 9.2 Definitions of working time call record and effective record

- Define the earliest call records before 8:00 and the latest call records after 20:00 as effective records because during 3:00–8:00 and 20:00–3:00 people are more likely to stay home.
- Count the effective record frequency of the different BSRs and get the most frequent BSR.
- If the frequency is more than 10 (at least 1 record per day), we set this BSR as home location for the user.

- For work location:

  - Count the frequency of the working time (9:00–18:00) call records in different BSRs and obtain the most frequent BSR.
  - For the BSR above, count the number of days that have at least one working time call record.
  - If the number is larger than 10, we set this BSR as work location for the user.

We get 10,790,048 call records from 12,846 users whose home location and work location can be identified by the above rules.

9.2.2 Latent Dirichlet Allocation Model

Topic model is a type of statistical model for discovering the abstract "topics" that occur in a collection of documents. LDA, introduced by Blei in 2003, is the most common topic model currently used for collections of discrete data. It was originally used for text analysis which can identify the latent topics for documents with a set of words [17, 18]. We can get topic distribution of each given document and

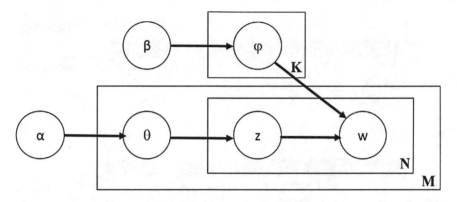

Fig. 9.3 Graphical models of latent Dirichlet allocation (LDA)

word distribution of each topic through word distribution of documents. Nowadays, the LDA model is widely used in the analysis of image, video, and so on. In this paper, LDA model is performed to find the latent mobility topic behind the location information sequence.

Figure 9.3 shows the generative process of the LDA model. Let α and β be the hyper parameters for Dirichlet document-topic distribution and topic-word distribution, respectively. θ is an $M \times K$ matrix of topic proportions for the K topics drawn from Dirichlet(α) and φ is a $V \times K$ matrix of distribution over vocabulary for the K topics drawn from Dirichlet(β). The topic assignments for a given document are $Z = (Z_1, Z_2, \ldots, Z_K)$ drawn from multinomial distribution with parameter θ. The words of the document are $W = (W_1, W_2, \ldots, W_N)$ drawn from multinomial distribution with parameter φ.

The main objective of LDA is to obtain topic distribution of every given document and word distribution of every topic.

Parameter perplexity is used to acquire the best latent topic number of the model [18]. The perplexity, used by convention in language modeling, is monotonically decreasing in the likelihood of the test data, and is algebraically equivalent to the inverse of the geometric mean per-word likelihood. A lower perplexity score indicates better generalization performance. More formally, for a test set of M documents, the perplexity is:

$$\text{perplexity}\,(D_{\text{test}}) = \exp\left\{-\frac{\sum_{d=1}^{M} \log p\,(W_d)}{\sum_{d=1}^{M} N_d}\right\} \tag{9.1}$$

where d is document, W_d is a sequence of word in document d, and N_d is the number of words in document d.

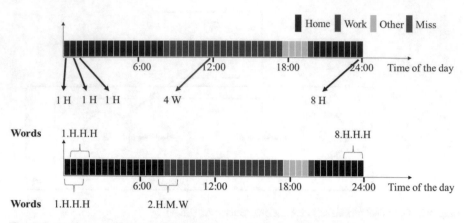

Fig. 9.4 Words generation. One day (one document) consists of 46 words

9.2.3 Bag of Location Sequences

We divide each day into 48 timeslots (Farrahi and Gatica-Perez [15]), where every timeslot is 30 min and at the same time, location information for every timeslots (H—Home, W—Work, O—Other, and M—No record) is also labeled. Then, every three consecutive timeslots is considered to be a sequence. Lastly, we add the coarse-grain timeslots label to every sequence (one day can be divided into eight coarse-grain timeslots, (1) 1–7 am, (2) 7–9 am, (3) 9–11 am, (4) 11 am–2 pm, (5) 2–5 pm, (6) 5–7 pm, (7) 7–9 pm, and (8) 9–12 pm). Thus, the words of the LDA model have been generated (see details in Fig. 9.4). By calculating the words frequency vector of the users, we obtain the bag of location sequences, which can be the input of the LDA model.

9.2.4 Clustering Algorithm: Affinity Propagation

Affinity propagation (AP) adopts the measures of similarity between pairs of data points to determine the cluster. The number of clusters need not be pre-specified and all the data points are thought to be the cluster centers in the algorithm, named "exemplars." Real-valued messages are exchanged between data points until a high-quality set of exemplars and corresponding clusters occur. Affinity propagation found clusters with much lower error than other methods [19].

In this study, we use AP clustering to extract mobility pattern. The input feature is the topic distribution generated by LDA model. The similarity of two mobility topic distribution is measured by Euclidean distance.

9.3   Results and Discussions

For every user, calculate the average number of effective timeslots (timeslots with call records) per day. Figure 9.5 is the user distribution of effective timeslots number. To describe the location information better, we use parameter q, which means the fraction of noneffective timeslots (timeslots with no call records) [10], and select days' of users with $q < 0.8$. Furthermore, we select the days in which the location information of 48 timeslots can be totally and clearly identified with H, W, O, and M (H—Home, W—Work, O—Other, and M—No record). After cleaning, we use 3371 days' records of 287 users on weekday and 1014 days' records of 275 users on weekend.

9.3.1   Finding the Best Latent Topic Number of the Model

The perplexity of different number of topics is showed in Fig. 9.6. The perplexity tends to be stable when k reaches 80 for weekday and 35 for weekend. So, we set k equal to 80 and 35 for weekday and weekend, respectively.

9.3.2   Model Results

We choose $k = 80$ for weekday and $k = 35$ for weekend to calculate the results of the LDA model, and we obtain the daily probability distribution matrix of topics and probability distribution matrix of words for every topic (part of results is shown

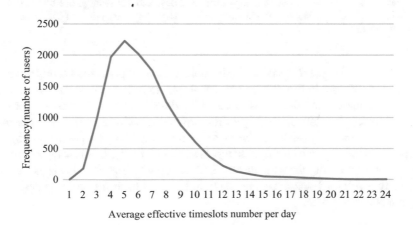

Fig. 9.5 User distribution of effective timeslots number. Number of effective timeslots for most users is less than 10 because the sparsity of the mobile phone record data

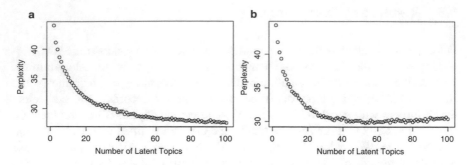

Fig. 9.6 Perplexity of different topics. (**a**) represents weekday and (**b**) as weekend. Perplexity decreases and then gradually becomes stable with the increase of topic number

Table 9.2 Some results of weekday by LDA model

Day	Topic						
	1	2	3	4	5	6	7
1	0.0065	0.0065	0.0065	0.0065	0.0065	0.0065	0.0378
2	0.0065	0.0065	0.0065	0.0065	0.0065	0.0065	0.0065
3	0.0065	0.0065	0.0065	0.0065	0.0065	0.0065	0.0378
4	0.0065	0.0169	0.0065	0.0065	0.0065	0.0065	0.0898
5	0.0065	0.0065	0.0065	0.0065	0.0065	0.0065	0.0169
Topic	Word						
	1.H.H.H	1.H.H.M	1.H.H.O	1.H.H.W	1.H.M.H	1.H.M.M	1.H.M.O
1	0.0001	0.0001	0.0001	0.0001	0.0001	0.0001	0.0001
2	0.0001	0.0001	0.0001	0.0001	0.0001	0.0001	0.0001
3	0.0001	0.0006	0.0001	0.0006	0.0001	0.0012	0.0001
4	0.0001	0.0001	0.0001	0.0001	0.0001	0.0001	0.0001
5	0.7421	0.1077	0.0000	0.0000	0.0000	0.0812	0.0008

For weekday, we get distributions of 80 topics for 3371 days and distributions of 512 words for 80 topics. For weekend, distributions of 35 topics for 1014 days and distributions of 512 words for 35 topics are obtained

in Table 9.2). We observe that a single topic cannot explain the mobility well in our case. As shown in Fig. 9.7, top 5 probability topics are chosen from the topic distribution matrix and the top 10 probability days' location information for each of the 5 topics are plotted. The higher the probability of a topic in a day, the more evident is the mobility related to the topic for this day. Thus, we get the highlighted 5 topics and the days which show the topics most obviously. In this way, we are able to know what mobility the topic means. The reason why results of the topics are not satisfactory will be discussed in next section. Finally, clustering has been performed to get a better outcome.

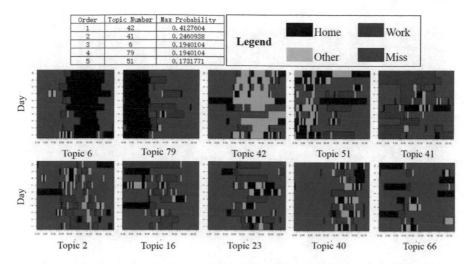

Fig. 9.7 Top 10 probability days' mobility for some topics. Different colors mean different locations of users. The horizontal axis is time of the day and vertical axis shows different days. Each row of the figure represents 1 day's mobility and by this way the mobility of different days can be specifically described. The five figures above are the top five probability topics among the topic distribution matrix and the five figures below are some topics without expectation. Topic 6 and topic 79 are mobility about home location but at different times, topic 42 and topic 51 are about other locations; however, obvious regularity cannot be captured from other topics in this figure

9.3.3 Cluster Results and Analysis

Before clustering the results of the LDA model, two issues ought to be addressed. The reason for expending a lot of effort on clustering the LDA model results instead of using the topics directly to explain mobility patterns and why we do not cluster the location information directly. For the first question, on the one hand, it is hard to model the mobility by single topic because the mobility pattern of the day is represented by the distribution of all the topics even though some day obviously shows one topic and it is impossible to explain all the days' patterns by topics directly. On the other hand, with the sparsity of the CRD (showed in Fig. 9.5), the topics about "M" (missing location information) will account for a large part of the results (e.g., blue part of topic 16, 23, and 40 showed in Fig. 9.7) and the significant information (topics about "H," "W," and "O") may easily be ignored if we just consider single topic. For the second question, performing a cluster first does not yield satisfactory results. The main reason can be attributed to the sparsity of our data. When we calculate the distance matrix of our data, a lot of "M" will adversely impact the final results of the cluster and thus weakens the real information of interest to us ("H," "W," and "O"). But, by considering the similarity of the distribution of the topics, this problem can be addressed. Some results by clustering the location information directly are shown in Fig. 9.8 using the same cluster algorithm and the same data.

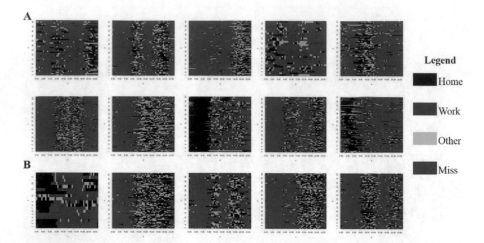

Fig. 9.8 Results of clustering the location information directly. (**a**) is for weekday and (**b**) is for weekend. Details about the figure are shown in Fig. 9.7. Much noise makes it difficult to classify mobility patterns in this way

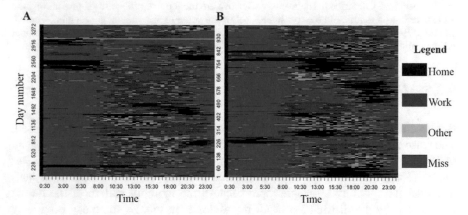

Fig. 9.9 Cluster results of affinity propagation. (**a**) is for weekday and (**b**) is for weekend. Details about the figure are showed in Fig. 9.7. We can see several main patterns in this figure and the different density of the color shows the different mobility on weekday and weekend

We use affinity propagation (*19*) to cluster the topic distribution matrix and we get 25 clusters for weekday data and 17 clusters for weekend data. The location information of all the users are plotted in Fig. 9.9 and users in the same cluster are put together.

Figure 9.9 shows the different classification of the mobility patterns of the days on weekday and weekend. We can easily get the location information during daytime because people usually make calls during this time. But, from 0:00 to 6:00, people make few calls. Concerning this situation, we label the timeslots between two consecutive calls with home or work location when the interval between two consecutive calls at the same location (home or work) is less than 8 h. For other locations, we use 4 h. But, not all users make calls before they go to sleep and after they wake up at home location and this is the reason why there still are a lot of blue grids in the figure. Generally speaking, activities of users at late night are few, and if they have some entertainment or work behaviors at that time, the probability they make calls is higher than usual and we are more likely to capture the location information. Above all, the missing location information does not affect the results too much when we extract the main travel behaviors of the users.

Left part of the Fig. 9.9 is about weekday and users of our data are commute users whose home location and work location can be identified, so the cluster result shows high density for working activities. Most of the users are working between 9 am and 6 pm, which is in accordance with the result of Shenzhen travel survey data in 2010, and some users do not finish their work until 9 pm. In [20], we also conclude that the evening peak extends to 3–4 h in Shenzhen. Few users work on the night shift. And some users are not at the work spots all day, for example, jobs like watchman may work for some days and have some days off. Right part of the Fig. 9.9 is for weekend and we can see lower density of working than weekday. But, some users still work on the weekends and even have higher work intensity in some days. This situation is normal for some jobs like services. Some users stay home all day on the weekends and most users do not leave far away from home (because of the black and green stripes). Few users go out all day, they may go to some remote places to spend their weekends.

Results of different clusters are plotted, respectively, in Fig. 9.10. We can see a very significant regularity on the whole, but there still remains a lot of noise from the micro perspective. Because of the limitation of the CRD, the same travel behavior of a user may show different results by mobile phone data, and sometimes even two calls in the same location may be recorded in different BSRs due to the signal intensity of the base station. This limitation can be showed by the jagged shape of location switching boundaries in the figure. As mentioned before, sparsity of the CRD may result in the loss of some location information of the users, which is another limitation. Hence, the results cannot be used to assess microlevel locational analysis but can be used to aggregate patterns of travel. We have changed the parameters of the AP algorithm to put all the information of the data into consideration. However, too many clusters (e.g., weekday data has 190 clusters) makes our result too specialized to make sense. Regarding the noise as the part of results will make it difficult to capture the real and main mobility patterns of the users. Thus, like the fuzzy processing showed in Fig. 9.11, the mobility patterns are more distinct when we make the streaks fuzzy and negative effects of the noise can be reduced.

In order to better describe the characteristics of each cluster, we regard repetitive behavior patterns of most days of the cluster as the type of mobility pattern for this cluster. If $\varphi_l^i(j)$ represents the count of location label l (H, W, O, M) on the day of number i in the cluster number n during timeslot j ($\varphi_l^i(j) = 0, 1$), then

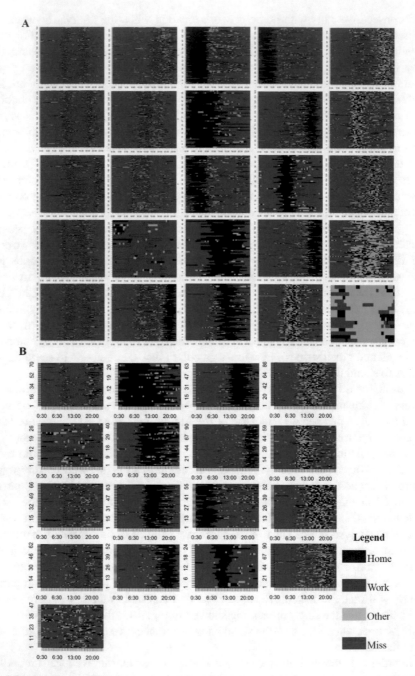

Fig. 9.10 Results of different clusters. (**a**) is for weekday and (**b**) is for weekend. Details about the figure are showed in Fig. 9.7. There are 25 clusters on weekday and 17 clusters on weekend

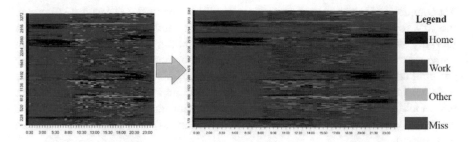

Fig. 9.11 Fuzzy processing of the figure (for weekday). After fuzzy processing, the color becomes more regular and the mobility patterns are more distinct

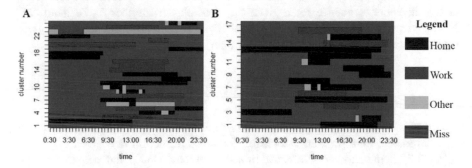

Fig. 9.12 Mobility feature of different clusters. (**a**) is for weekday and (**b**) is for weekend. Different colors represent different locations for users. The horizontal axis is time of the day and vertical axis shows different clusters. Each row of the figure is the mobility feature of one cluster and by this way mobility features of all clusters can be clearly described. It is worth noting that the mobility of cluster 17 on weekend does not have any location information because the cluster shows less regularity (we can see the last figure of Fig. 9.10b)

$$f_n(j) = \arg \max_l \sum_i^{D_n} \varphi_l^i(j) \tag{9.2}$$

where $f_n(j)$ is the assigned characteristic location label for timeslot j of cluster number n and D_n is the total day number of cluster n. Thus, we can get the representative mobility feature of every cluster in Fig. 9.12 and we can know what pattern the cluster represents. It can be found that people's main mobility is very regular and focuses on several patterns which can be easily explained by our life experience. Particularly, the mobility patterns are just generated by CRD and we have no idea about the location information for the timeslots with no call. As mentioned above, most of the timeslots with unknown location information are in 0:00–6:00, which is the time for sleeping and with low activity intensity, but if they have some entertainment or work behaviors at this time, it also has a greater possibility to make a call. So, we believe that our results can capture the main mobility patterns of the users. It needs to be noted that all the classes are not completely independent. For example, as illustrated in Fig. 9.10a-1-4 (Figure in the first row and fourth column of the Fig. 9.10a) and Fig. 9.10a-2-4, the main mobility feature of the two clusters is at home before 6:30 (cluster 22) and at home after 17:30 (cluster 12). These two daily behavior patterns of weekdays are not contradicting. As pointed out before, the same behavior pattern may generate different results because of limitation of the CRD and these two patterns maybe two mobility pieces of one behavior captured by our data. Through extracting the behavior characteristics of several days, the complete mobility patterns of users can be acquired.

We use the cluster feature to explain the mobility and the parameter φ is defined to describe the accuracy of the results. Parameter φ is used to measure the similarity between real mobility and the mobility of representative cluster feature. If L_{ij} is the timeslots' sequence for a given day i of cluster number j (the label "M" is not effective location information), and F_j is timeslots' sequence for a given representative feature of cluster number j, then

$$\varphi_j = \frac{1}{D_j} \sum_{i=1}^{D_j} \frac{S\left(L_{ij} \cap F_j\right)}{N\left(L_{ij} \cap F_j\right)} \tag{9.3}$$

where φ_j is the accuracy of cluster j, and $L_{ij} \cap F_j$ is the timeslot sequence both have the real location information and cluster feature information in a given day i, and $N(L_{ij} \cap F_j)$ is the number of effective timeslots in the $L_{ij} \cap F_j$, and $S(L_{ij} \cap F_j)$ is the number of the timeslots which have the same location information in the $L_{ij} \cap F_j$ about real and cluster situation, and D_j is the days' number of cluster j. The results of φ_j are showed in Fig. 9.13. The mean value of φ is 0.841 for weekday and 0.837 for weekend. Almost all clusters have reached high accuracy. Low accuracy of some

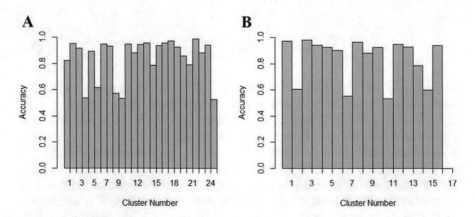

Fig. 9.13 Accuracy of cluster feature. (**a**) is for weekday and (**b**) is for weekend. Almost all clusters have reached high accuracy. Low accuracy of some clusters can be subject to the fluctuation of home and other places (black and green stripes in Fig. 9.10)

clusters can be subject to the fluctuation of home and other places (black and green stripes in Fig. 9.10). On the whole, our cluster results can show one's mobility to a great degree.

9.4 Conclusions

In this study, we simplify users' travel destination by home, work, and other to describe their mobility. By clustering the output of the LDA model, we reduce the negative impact of missing information to the cluster results and identify the mobility patterns of the mobile phone users (25 classes on weekday and 17 classes on weekend). The results are explainable and consistent with our life experience.

Locating users using CRD has its own advantages and limitations. On the one hand, it has a large sample which can be accessed continuously for a very long time and with little deviation (almost everyone has a mobile phone and most traffic modes can be covered, while people have various usage habits). On the other hand, localization error and data sparsity are the main limitation of the methodology, and we are unable to obtain the social attributes of the users. Hence, we would get stuck if we excessively pursuit the accurate result. It is very likely to treat the location error as a part of results, which will make it difficult to capture the real and main mobility patterns of the users. For data sparsity, we just consider the mobility which is shown by the CRD and by combining the behavior characteristics of several days, the relatively complete mobility patterns of users can be acquired. For localization noise, we use LDA model, AP algorithm, and representative feature extraction to capture the main mobility patterns of the users, finding that almost all the mobility of users gathered in several classes.

Many researches indicate that humans follow simple reproducible patterns with high predictability [10, 21], and our results support this conclusion from the perspective of the activity sequence based on home and work location. For commuter, home and work make up a high proportion of daily mobility and show different density on weekday and weekend.

Through learning the mobility of the users, we can predict the future travel behaviors for various users. There still remain doubts about how to choose classifier and how to deal with the situation that all the classes are not completely independent with each other. Therefore, multi-label classification will be an alternative choice in our future research.

Acknowledgments This research was sponsored by National Natural Science Foundation of China (71171147) and Fundamental Research Funds for the Central Universities.

References

1. Z. Wang, S.Y. He, Y. Leung, Applying mobile phone data to travel behaviour research: a literature review. Travel Behav. Soc. 11 (2018)
2. S. Lu, Z. Fang, X. Zhang, S.-L. Shaw, L. Yin, Z. Zhao, X. Yang, Understanding the representativeness of mobile phone location data in characterizing human mobility indicators. ISPRS Int. J. Geo-Inf. **6**(1), 7 (2017)
3. S. Jiang, J. Ferreira, M.C. Gonzales, Activity-based human mobility patterns inferred from mobile phone data: a case study of Singapore. IEEE Trans. Big Data **3**, 208 (2016)
4. J. White, I. Wells, Extracting origin destination information from mobile phone data, in *Eleventh International Conference on Road Transport Information and Control* (IET, 2002)
5. M. Friedrich, P. Jehlicka, T. Otterstätter, J. Schlaich, M. Friedrich, P. Jehlicka, T. Otterstätter, J. Schlaich, Monitoring travel behaviour and service quality in transport networks with floating phone data, in *Proceedings of the 4th International Symposium Networks for Mobility* (Stuttgart University, Stuttgart, 2008), pp. 1–7
6. A.J. Lee, Y.-A. Chen, W.-C. Ip, Mining frequent trajectory patterns in spatial–temporal databases. Inf. Sci. **179**(13), 2218–2231 (2009)
7. A.A. Shaw, N. Gopalan, Frequent pattern mining of trajectory coordinates using Apriori algorithm. Int. J. Comput. Appl. **22**(9), 1 (2011)
8. S. Abraham, P.S. Lal, Spatio-temporal similarity of network-constrained moving object trajectories using sequence alignment of travel locations. Transp. Res. C **23**, 109–123 (2012)
9. D.-H. Shih, M.-H. Shih, D.C. Yen, J.-H. Hsu, Personal mobility pattern mining and anomaly detection in the GPS era. Cartogr. Geogr. Inf. Sci. **43**(1), 55–67 (2016)
10. C. Song, Z. Qu, N. Blumm, A.-L. Barabási, Limits of predictability in human mobility. Science **327**(5968), 1018–1021 (2010)
11. R. Ahas, S. Silm, O. Järv, E. Saluveer, M. Tiru, Using mobile positioning data to model locations meaningful to users of mobile phones. J. Urban Technol. **17**(1), 3–27 (2010)
12. S. Phithakkitnukoon, T. Horanont, G. Di Lorenzo, R. Shibasaki, C. Ratti, Activity-aware map: Identifying human daily activity pattern using mobile phone data, in *International Workshop on Human Behavior Understanding* (Springer, Berlin, 2010), pp. 14–25
13. S. Hasan, S.V. Ukkusuri, Urban activity pattern classification using topic models from online geo-location data. Transp. Res. C **44**, 363–381 (2014)
14. K.S. Kung, K. Greco, S. Sobolevsky, C. Ratti, Exploring universal patterns in human home-work commuting from mobile phone data. PLoS One **9**(6), e96180 (2014)
15. K. Farrahi, D. Gatica-Perez, Discovering routines from large-scale human locations using probabilistic topic models. ACM Trans. Intell. Syst. Technol. **2**(1), 3 (2011)
16. J.E. Spinney, Mobile positioning and LBS applications. Geography **88**, 256–265 (2003)
17. D.M. Blei, Probabilistic topic models. Commun. ACM **55**(4), 77–84 (2012)
18. D.M. Blei, A.Y. Ng, M.I. Jordan, Latent Dirichlet allocation. J. Mach. Learn. Res. **3**, 993–1022 (2003)
19. B.J. Frey, D. Dueck, Clustering by passing messages between data points. Science **315**(5814), 972–976 (2007)
20. C. Yang, Y.L. Zhang, F. Zhang, Commute feature analysis based on mobile phone data: case for Shenzhen. Urban Transp. China **14**(1), 30–36 (2016.) (in Chinese)
21. M.C. Gonzalez, C.A. Hidalgo, A.-L. Barabasi, Understanding individual human mobility patterns. Nature **453**(7196), 779–782 (2008)

Index

Printed in the United States
By Bookmasters